THE AMBER FOREST

The
AMBER FOREST

A Reconstruction of a Vanished World

GEORGE POINAR, JR., AND
ROBERTA POINAR

WITH PHOTOGRAPHS AND DRAWINGS
BY THE AUTHORS

PRINCETON UNIVERSITY PRESS, PRINCETON, NEW JERSEY

Library of Congress Cataloging-in-Publication Data
Poinar, George O.
The amber forest: a reconstruction of a vanished world/
George Poinar, Jr., and Roberta Poinar:
photographs and drawings by the authors.
p. cm.
Includes bibliographical references and index.
ISBN 0-691-02888-5 (cloth: alk. paper)
1. Amber fossils—Dominican Republic.   2. Amber—Dominican Republic.
3. Paleoecology—Dominican Republic.   I. Poinar, Roberta. II. Title.
QE 742.P64   1999
560'.45'097293——dc21   98-35388

This book has been composed in Palatino.

The paper used in this publication meets the
minimum requirements of
ANSI/NISO Z39.48-1992 (R1997)
(Permanence of Paper)

http://pup.princeton.edu

Printed in the United States of America

10 9 8 7 6 5 4 3 2

*This book is dedicated to
our children and grandchildren
and to the inquisitive child within us all
who can be delighted by the
wonders of the natural world*

From time to time, a few truths are revealed, tiny pieces of the vast mosaic of things. Better to divulge the discovery, however humble it be. Others will come who, also gathering a few fragments, will assemble the whole into a picture ever growing larger but ever notched by the unknown.

J. Henri Fabri, from vol. 10, *Souvenirs entomologiques*, 1923

Contents

Preface

MILLIONS OF years ago, a forest flourished on the island now known as Hispaniola. One of the canopy trees, the algarrobo, was unique in that it exuded copious amounts of resin which acted as a sticky trap, catching and holding invertebrates, small vertebrates, plant parts, and other evidence of the wildlife inhabiting that ancient land. This resin entombed the biota and preserved it, ultimately undergoing a fossilization process that turned it into amber. The animals and plants conserved in amber from the Dominican Republic represent one of the world's most complete fossil records of terrestrial life in a tropical region. The sheer variety and magnitude of this particular deposit allow one to garner some perspective of this bygone forest.

Here we travel back in time to record the life and activities of these extinct organisms by using their remains to reconstruct many aspects of the natural environment. The present work shows how clues from various fossils can be used in the paleoreconstruction of an ancient forest, revealing the environment, the climate, the natural history of many life forms, and even the interactions among organisms.

A Note on the Illustrations

This book contains three types of illustrations: black and white photos, a color-plate section of selected photos from the black and white images, and pencil drawings reconstructing the probable environment of the flora and fauna. In the drawing captions, we indicate which photos represent or correspond to the objects in the drawings.

Captions for the color photos are the same as for the corresponding black and white photos. A (c) following the figure caption indicates that the photo is also in the color section.

Acknowledgments

WE WOULD like to thank the following individuals for various types of assistance provided during the preparation of this book, especially regarding identification, biological data and pertinent literature: P. Adams, W. Allmond, N. Anderson, N. Møller Andersen, G. Ball, C. Baroni Urbani, J. Baxter, D. Benzing, S. De Bortoli, E. Both, A. Boucot, M. Bourell, D. Bright, A. Brown, W. R. Buck, L. Caltagirone, M. Capriles, D. Cavanaugh, D. Chandler, J. Chemsack, Y. Coineau, J. Cokendolpher, T. Columbus, P. Craig, R. Crowson, C. Dietrich, P. DeVries, R. Disney, J. Doyen, J. Frahm, H. Frank, J. Gelhaus, S. Gradstein, W. Grogan, J. Halaj, P. Hammond, M. Hansen, L. Kimsey, D. Kistner, G. Krantz, M. Kogan, L. Knutson, J. LaBonte, J. Lattin, L. LeSage, R. Lewis, J. Liebherr, P. Macafferty, E. and R. Maroni, R. Mason, M. Meinander, L. Mendes, K. Merrifield, C. Michener, J. Miller, E. Mockford, A. Moldenke, D. Nelson, A. Norrbom, D. Norton, S. O'Keefe, C. W. O'Brien, L. O'Brien, G. Parsons, N. Penny, E. Petersen, S. Podenas, M. Prentice, G. Pritchard, F. Radowsky, A. P. Rasnitsyn, G. Rogers, B. Roth, R. Roughley, J. Santiago-Blay, A. Scarbrough, T. Schowalter, A. Smith, L. Stange, B. Stephen, J. Strother, H. Sturm, C. Thompson, P. Ward, Q. Wheeler, T. Zavortink, and T. Zug. Grateful appreciation is extended to the many amber dealers, especially Jim Work, Dan McAuley, and Jake Brodzinsky, who sent us interesting material to examine. We also thank the private collectors who allowed us to photograph their specimens.

All of the specimens depicted here are deposited in the Poinar amber collection maintained at Oregon State University, except for those in the following figures: 48, 53, 76, 93, 126, 127, 129, 135, 143, 144, 160, 161—M collection; 84, 163—Smithsonian collection; 121, 122—Work collection; 137—Cardoen collection; 77—Truman collection; 164—Costa Amber Museum; 50—Lyman Entomological Museum.

Prologue

I LIFTED to the window a nugget of golden Dominican amber entombing a small stingless bee. The sunlight infused it and illuminated the bee caught forever in flight—gossamer wings outstretched and perfectly preserved down to the last hair. Stark eyes appeared to be gazing at me. I contemplated this lustrous burial chamber and thought how wondrous it would be if we could see what this insect had beheld in its lifetime. Would the vistas of just one day be sufficient to reveal the wonders of life millions of years ago? What was that last fateful day like? And what events had taken place in the eras before this specimen arrived in my hand?

One surmises that the bee was active in the dim light of early morning. She and her sisters gathered in the busy colony before beginning their various tasks. The young workers left for the nursery to attend the developing larvae. Older members flew out into the forest to collect pollen and nectar. The chore reserved for the aged bees was collecting the sticky resin utilized in nest construction from algarrobo trees. Our bee was among the resin gatherers. Off she skimmed with her companions, winding through the shadowy, towering amber forest, dodging the vines and lianas, avoiding tree trunks where hungry lizards lurked, and finally landing on the bark of an algarrobo near a large, yellow, viscous resin flow.

She scanned the surroundings, always on the lookout for hungry, sinister creatures that lurked in ambush—especially one well-adapted predator, the resin bug, a large, hairy-legged creature endowed with a huge beak that could easily penetrate the body and suck out the blood of an unwary bee. The resin bug's habit of coating the front legs and body parts with resin was repulsive, though effective in ensnaring prey. Only one swipe of the powerful front legs could pin the hapless victim long enough for the hypodermic-like beak to rip through the body wall.

Scenes from the amber forest as viewed by a stingless bee in the last few hours of its life.

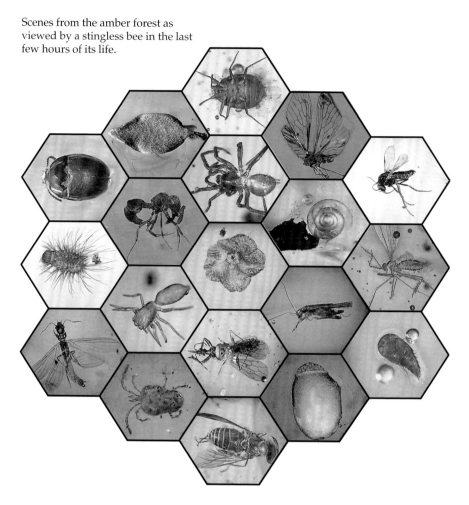

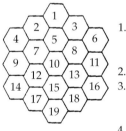

1. Scale insect (Coccoidea: Homoptera)
2. Leaf
3. Planthopper (Derbidae: Homoptera).
4. Darkling beetle (*Liodema phalacroides*: Tenebrionidae: Coleoptera).
5. Spider (Araneae: Arachnida).
6. Biting midge (Ceratopogonidae: Diptera).
7. Ant (Formicidae: Hymenoptera).
8. Snail (*Helicinidae:* Gastropoda).

9. Beetle larva (Coleoptera).
10. Flower
11. Gall midge (Cecidomyiidae: Diptera).
12. Spider (Araneae: Arachnida).
13. Winged bush cricket (Trigonidiidae: Grylloptera).
14. Web spinner (*Mesembia:* Anisembiidae: Embioptera).
15. Gnat bug (Enicocephalidae: Hemiptera).
16. Termite-nest beetle (Trichopseniinae: Staphylinidae: Coleoptera).
17. Velvet mite (Thrombidiformes: Acari).
18. Egg of stick insect (Phasmatidae: Phasmida).
19. Horsefly (Tabanidae: Diptera).

The bee's compound eyes registered a kaleidoscopic image of the resin flow. Trapped within the vitelline pool lay other small arthropods, plant debris, and detritus. Not discerning any dreaded enemies, the bee began the painstaking job of removing small samples of resin with her mouth, coating them with saliva, and then attaching them to the hind legs in the form of little round balls. This exercise involved concentration and diligent work to prevent getting entrapped in the adhesive deposit. Finished at last, she was ready to return to the colony.

The attack came with lightning speed. Only a hazy brown blur was detected as the resin bug thrust its front legs toward her. Acting on impulse, she retreated from the lunging bug, but in her frantic attempt to avoid the predator she flew directly into the sticky trap. Almost instantly, waves of thick fluid enveloped her.

A few feeble attempts were all that could be mustered to extricate herself from that tenacious snare. In spite of valiant efforts, death came in seconds as viscous liquid seeped over the breathing pores, wrapping a mantle of gold around its victim. The sun shimmered on the silent insect cradled in the glistening tomb on the algarrobo trunk. As a gentle breeze wafted through the leaves, a fine layer of dust drifted over the surface of the elixir. Suspended in a motionless world, the entombed bee endured, the balls destined for the hive still attached to her outstretched legs. Almost immediately, chemical changes began to take place. Sugars in the sap pulled the moisture from the insect's tissues while other chemicals infiltrated its cells. Bacteria carried in the gut initiated spore development in response to the adverse conditions. The next day, the xanthous domain of the bee was covered by a series of subsequent flows. Changes were also occurring in the resin itself as exposure to sunlight and oxygen caused the bonds between molecules to strengthen.

Days grew into weeks and weeks into months. The tacky mass became harder but still clung to the bark of the tree. Finally, perhaps during a storm, the fossilized resin broke loose and came crashing down to the ground, lodging in a small crevice at the base of the tree. The material was now in the copal stage and no longer sticky. The chemical processes of polymerization and molecular cross-bonding would continue over the next few million years or so until the hardened resin, or copal, acquired the properties of amber. Year after year, more debris, leaves, and twigs collected on top of the hardening copal. Wind, rain, and microorganisms degraded the accumulated plant litter, but had no effect on our entombed bee. Eventually the mighty algarrobo crashed to the forest floor in a storm and added still another layer to the detritus covering the fossil.

For millennia all was tranquil. Ultimately a tempest fueled by a hurricane drove torrential rains across the bee's graveyard, washed away the accumulated organic matter, and exposed the fossil once again. Along with other fossilized pieces and plant debris, the specimen was washed by rivulets of floodwaters into raging streams, transported into a low-lying delta and shrouded with silt.

The sea level began to change and salt water slowly inundated the area, submerging our entombed bee. Ocean waves dislodged the silt and exposed the amber. Around the fossil now lived an array of sea creatures. Crabs scampered over the resin graveyard, fish swam up and nudged amber nuggets, barnacles selected larger pieces for resting sites, and in heavy storms the fossilized resin was tumbled repeatedly by the currents. Among the waves dwelled a myriad of microscopic, shelled animals called foraminifera and coccoliths. When they died, their minute shells settled to the bottom and, together with bits of sediment, slowly covered the precious fossil on the sea floor. No longer would a diffuse light illuminate our bee, not for millions of years.

Meanwhile, the mass of earth containing the amber alluvium drifted farther into the Caribbean Sea, having already moved from its original position between North and South America.

The advance of this island mass was gradual, and those that once lived in that extinct silva were indifferent to the earthquakes that periodically shook the terrain.

Eons passed and the sediment covering the amber now hardened into rock. Ultimately the strata containing the entombed creatures were subjected to still another natural force, that of mountain formation. Ponderously, those rocks that had been forged in the sea were elevated above the water. In some areas, the layers shifted and fractured, sometimes crazing the pieces within. The bee was fortunate to be in a section of rock that was uplifted in its entirety. Towering skyward, the strata were folded into mountains. The upper boundaries, whose history could be told by the marine microorganisms in their matrices, were worn by the wind, washed by the rain, and eventually transformed into soil. Ultimately another forest became established, sending its roots down into the new loam that covered the bejeweled rocks, one quite dissimilar from the original forest.

One day, the silent graveyard, now lodged in a layer of marine rocks in the mountains of the northern Dominican Republic, was disturbed by some minor shock waves. These jolts continued for several days, eventually becoming more intense. The sediment containing the amber had been discovered and was being laboriously removed. Suddenly, the sarcophagus was jarred as a hammer cracked open the rock in which it was inhumed. Together with other pieces that had been dislodged from the rock, the bee and its tomb toppled to the floor of the mine. In the flickering candlelight, she was picked up and placed into a bag. Later, the day's take was sorted on a rickety wooden table where dark eyes scrutinized each piece intently, selecting those with fossils to be polished and sold and thereby continuing the process that brought this bee to my collection.

Placing the bee back into a box brimming with hundreds of other specimens, I ran my hand through this assorted group and reflected how each contained a different individual with its own story to tell. Could a compendium of these clues be used in a paleoreconstruction of the forest that existed millions of years ago? Every specimen represented a piece of the amber forest jigsaw puzzle. Describing an extinct tropical forest from amber fossils would be a unique approach in paleoecology and could reveal how this past realm differed from any existing today.

THE AMBER FOREST

1

Introduction

From Where and Whence

Step back in time and explore with us a primeval forest that flourished some 15–45 million years ago and then disappeared, leaving testimony of its existence in amber from the Dominican Republic. The description of this extinct forest will be extrapolated from select fossils recovered over the past 20 years. Before our exploration of the ancient forest begins, some important facts are needed to gain a perspective of the events that occurred so long ago. These facts are analogous to the knowledge an explorer of today would gather before embarking on a trip—a bird's-eye view of the history and environment of a foreign country. Of course the time frame is quite different.

When asked "Where are we going?" the answer is more complex than just to the Dominican Republic or the Caribbean island of Hispaniola. We are traveling back to witness events that commenced during the age of the dinosaurs upon a mutable, shifting earth surface conveying an island across the sea on a journey that took millions of years. So we begin with the question of when Hispaniola first appeared in the earth's history. There are almost as many theories about the origins of the Greater Antilles and past earth movements in the Caribbean as there are scientists investigating this question.[1] Although the following scenario agrees in principle with most of these theories, some of the dates presented here were obtained from information based in part from amber fossils.

What are now the islands of the Greater Antilles (Cuba, Puerto Rico, Jamaica, and Hispaniola) first emerged as molten magma spewing from volcanic activity along a marine shelf between North and South America. This probably occurred some 100 million years ago in the Cretaceous period when the dinosaurs ruled the earth.[2] For millennia, this shelf remained beneath the sea but moved slowly east toward the Americas. It lacked the mass to breach the waves until, many millions of years later, subsequent lava flows pushed the rock above the sea and created a separate span of land roughly across where Central America is located today. This event, which connected North and South America, occurred near the end of the Cretaceous period, during which the earth sustained a cycle of extensive earthquakes and volcanic eruptions.

The newly formed terrain, known as the Proto-Greater Antilles, emerged barren and sterile about 65 million years ago. As the lava cooled, organisms began to arrive. First, the algae came as spores that were fortunate enough to land in moist crevices on the lava and germinate—thus starting a succession of life that in time would culminate in a mature forest. Soon after the algae were established, other spore-forming plants such as moss, ferns, and lichens found favorable niches for growth. These plants, together with the elements, modified the rocky surface just enough for small pockets of soil to accumulate, forming a foothold for higher plants. Next appeared seed plants in the form of herbs and shrubs. It didn't take long for dispersing insects to discover the plants, followed by spiders and parasites to prey on the insects and so on until a complete ecosystem with reptiles, birds, and mammals appeared. Eventually, the Proto-Greater Antilles was covered with a forest full of life garnered from the adjacent land masses. Animals and plants, thriving in what is now southern Mexico, spread onto the young land from the north, and from the south came floral and faunal elements from South America. Included among the new plants were algarrobo trees, producing the resin that would eventually form the basis of this work. This ancient forest represented an early assimilation of biota from North and South America.

The Proto-Greater Antilles land mass containing the foundation of the amber forest rested on the Caribbean continental plate that broke away from its position between North and South America and started moving slowly eastward about 60 million years ago. During this ponderous journey, other life forms would periodically arrive in the forest, some from the air, such as insects, and some by over-water dispersal, such as reptiles. At times of low ocean levels, when the land mass butted up against shallow ridges, animals might have been able to board the shifting ark. Some of these animals undoubtedly were successful in establishing themselves, thus adding new elements to the amber forest. The tectonic history of this terrain was quite complex as it smashed into and slid around land masses and other drifting plates along its way. Despite an eventful journey, it arrived in the Caribbean about 25 million years ago, with the area of the Dominican Republic remaining emergent since its appearance some 65 million years ago.[3]

Some questions remain regarding the age of the amber. A number of mines are scattered throughout portions of the northern mountain range of the Dominican Republic known as the Cordillera Septentrional as well as in the eastern part of the country.[4] Deposits from these diverse locations may be from different time periods, but just how different is not known. Dating has been attempted by chemical analyses of the amber[5] as well as by an examination of marine microfossils that occur in the bedrock. The youngest times suggested are 15–20 million years, based on foraminifera fossils,[6] while the oldest proposed are 30–45 million years, based on coccolith fossils.[7] Could this forest have survived for a much longer period, like 60 million years? If algarroba trees were producing resin during that entire period, what happened to the amber formed during the intervals spanning 4–15 million years and 45–60 million years ago? If a continual supply of amber did exist, was it destroyed by natural forces or does it still remain somewhere, locked away deep in the bowels of the earth? Or was the resin production limited to a relatively short period during the lifetime of a unique algarroba tree that produced copious amounts? In spite of the uncertainties regarding the exact dates, the vast array of fossils allows us to complete our paleo-reconstruction of the ancient forest.

Most of the amber lies buried in the fog-shrouded peaks of the north. The ranges cradling these treasures are thought to have formed in recent times, perhaps between 25,000 and 10,000 years ago.[8] During this period of mountain genesis, the sedimentary beds of limestone and shale containing the amber were uplifted. Emergence is a very tortuous process involving the bending and breaking of rock layers. Thus the fossilized resin was subjected to tremendous shear forces, which undoubtedly destroyed vast amounts of it and left extensive fractures in the remaining material. No wonder it is rare to find a complete piece larger than a golf ball without any internal fractures.

The Taino Indians may have been the first people to note the presence of this gem and appreciate its beauty, although other indigenous cultures were present on the island of Hispaniola when the Tainos arrived. When Christopher Columbus came in the 15th century, the Tainos presented treasured pieces to him, but the Spaniards were more interested in gold, and amber fell into oblivion. It was "rediscovered" at the beginning of the twentieth century, first as jewelry and then for its scientifically valuable fossils, and now it provides a source of income to many.

Amber mines are little more than small, tortuous tunnels carved into the sides of the mountains, or sometimes pits sunk deep into the ground. Certainly there is no mine in the Dominican Republic as elaborate as the one shown in the movie *Jurassic Park*, where miners wearing hard hats and carrying pneumatic drills were depicted pushing car loads of fossilized resin along rails from deep within a huge mine. In reality, it is rare to be able to stand fully upright in an amber mine. The narrow tunnels, which can continue 600 feet into the side of the mountains, are only sufficient for crawling. During the rainy season, tunnels partially filled with standing water are frequently encountered, and miners must crawl half-submerged through the dark confining space to reach the treasure-laden veins. During wet weather, when the soil is saturated, cave-ins and landslides sometimes bury and take the lives of unfortunate miners. Workers are equipped with only the most basic tools—a hammer, chisel, candle, and sack. Flickering candles serve to illuminate the dark confines of the tunnel, producing barely enough glow to follow the gray layer of sedimentary rock encasing the amber. Days are spent lying supine, crouching or kneeling in that small space, chiseling away at the matrix rock and picking out the nuggets of fossilized resin that are exposed. Many small fragments fall out as the rock is dislodged, and valuable specimens are often lost or damaged.

At the end of an exhausting day, the miners return home and, filled with anticipation, spread out their finds in the last rays of the afternoon sunlight. Morsels with rare or spectacular fossils are infrequent and these are placed aside. Eventually all of the marketable items end up at a workshop where young boys will polish them. Standing in front of noisy machines that run from dawn to dusk, the polisher holds the amber against sandpaper, removing dirt and sharp irregular edges and smoothing the surface. The hot, humid air seethes with powder, thus many of the workers wear masks. If a fossil is present, an attempt will be made to get as close as possible without wasting too much of the sample, since this gem is often sold by the gram.

Specimens are then turned over to a middleman who shows them to a variety of buyers searching for the best price. The scientific value of amber is, of course, due to the types of life forms it contains. If a fossil is large or rare, it can be a significant source of income to the owner. A vertebrate such as a frog or lizard can bring 30–40 thousand dollars, and large, well-preserved scorpions can be sold for as much as 15 thousand. The demand for choice fossils is so intense that many are now bought directly from the miner's sorting table.

It is difficult to say how many hands various fossils pass through before they eventually reach paleontologists. Many valuable and scientifically important items end up in the hands of private collectors. Some people will allow researchers to examine their personal collections to document new finds and describe new species. However, in many cases, individuals are unaware of significant fossils in their possession or are reluctant to share them, with the result that many remain undiscovered. There is no doubt that the great majority of rare and scientifically important specimens are in private collections, however, since most museums can little afford the exorbitant costs of unique amber fossils.

Why Select Amber for Study?

What is so singular about amber that merits using it to reconstruct an ancient forest? If you examine other types of preservation such as compression fossils, you will find that most insects are two dimensional, uniformly colored, and lack fine morphological details. In contrast, amber fossils provide edifying clues to the past for the following reasons: first, they are three dimensional; second, their original color patterns are often preserved; and third, the matrix has protected them from much of the abuse that other fossils receive—thus the preservation is outstanding. In addition, the organisms usually expired so swiftly that many are revealed in almost lifelike circumstances. Their accompanying organisms also had little chance of escaping. Thus there is evidence of past symbiotic associations that are rarely preserved under other circumstances. Finally, these fossils encompass plant, invertebrate, and vertebrate remains, all of which are important in a study like this one.

In each photograph presented here, you will see a clear picture of a frozen moment in the amber forest when it existed on a land now known as the Dominican Republic. Earlier we described what our bee may have seen in the final few hours of its life. By fitting these candid snapshots together like pieces of a jigsaw puzzle, we can make a kaleidoscope of life in its environs. The elegance of these fossils adds to their general enjoyment and fascination. Portrayed will be ancient life forms from a realm that no longer exists in the Greater Antilles, nor anywhere else in the world today! Such extinct forms can answer questions about paleobiodiversity, evolution, and biogeography. They can also enlighten us about past climates, insect-plant interactions, and parasitic relationships.

Amber fossils will be used to describe the primeval forest in relation to tropical forests as they exist today. There is no precedent for employing this technique for reconstructing a past tropical ecosystem, since aside from fossilized resin, the remains of organisms in tropical forests are few and scattered. High decomposition rates and low fossilization potential account for the meager number of actual fossils. In addition, insufficient rock exposure would render finding them almost impossible. The little that has been written to date about extinct tropical forests is based mainly on pollen records.

How Biased Are Amber Samples?

Can a selection of amber fossils portray a true sampling of past life in an extinct domain? One cannot hope to obtain evidence of all life forms that existed in that ancient forest. The majority of creatures preserved are those that lived or foraged on the bark of trees, such as worker ants, bark lice, certain beetles, and small flies. Such forms became entangled in the sticky resin as they performed their daily activities. The next most common category of organisms found are winged insects that were flying through the forest and alighted or were blown by a gust of wind against some sap on an algarrobo tree. A few insects, like our stingless bee and predator bug, were attracted for other purposes. Such organisms give a fairly accurate view of life in and around the algarrobo tree. Other, larger fossils from this same habitat are uncommon due to the limiting size of the resin flow (a small flow might only partially entrap a large specimen, making its remains susceptible to decay) and because of the creature's strength and ability to extricate itself from the sticky trap. Then there are the exceptional inclusions such as vertebrate remains, scorpions, orchid bees, and others. But rare does not just mean big, since fleas and ticks are as infrequent as frogs and lizards. These uncommon animals appear only occasionally and became entrapped by providence while searching for prey or by some unexplained misfortune. Based on experience, the likelihood of a rarity materializing is one in every five thousand pieces. Appendix A at the back of the book shows the frequency of types of organisms found in some three thousand pieces of amber. Obviously ants are the most abundant group represented, followed by gall midges, with bark lice close behind. Less frequent fossils include centipedes and grasshoppers. The more specimens examined, the more variety discovered. Each treasure is a scientific gem instructive in learning more about this former biome, yet unfortunately many of these are being sold as jewelry.

As we discuss the natural history of the extinct animals in amber, we will draw heavily on the principle of behavioral fixity in the fossil record. This principle is an important concept in the reconstruction of ancient worlds and states that the behavior, ecology, and climatic preferences of fossil organisms will be similar to that found in their present-day descendants at the generic and often family level. This well-documented principle is very helpful in deciphering the behavior of ancient organisms.[9]

Thus it may be possible to indirectly infer the presence of a plant or animal that existed long ago by finding an organism that today is intimately associated with it. When we discover fig wasps, for example, we can be certain that fig trees existed even though we have not found their actual remains, since it is known that fig wasps can develop only inside the fruits of this tree.

Structure of Tropical Forests

Present-day tropical forests are immensely diverse and complex, teeming with an incredible array of plants and animals. Yet all tropical silva the world over share certain characteristics.[10] Their lush vegetation consists mostly of evergreens, grouped according to size and growth habits. The vegetation is dispersed in more or less recognizable layers or strata. Beginning with the top, the canopy includes the bulk of the vegetation in tropical forests. Viewed from above, this layer forms a leafy roof appearing as an almost continuous undulating expanse of various patterns and colors. Flowers and fruits reaching up toward the sun provide splashes of vivid hues against a canvas with an amazing variety of textures. High above the floor, the branches abound with the many animals that seek food and shelter in this bountiful layer. The height of the trees in the canopy layer depends upon the type of forest, but in the amber forest it probably varied from 80 to 130 feet. Poking through this canopy layer to imposing heights of 160 feet are the giants of the forest, the emergent trees. Directly beneath the canopy is the subcanopy layer, which includes trees ranging from 40 to 80 feet. Underlying these layers, wherever patches of sunlight filter down into clearings or along the banks of wide river beds, is the understory layer, with residents growing from 20 to 40 feet. Among the understory vegetation, the inhabitants find refuge from the sun or rain. Still lower in the almost continuous shadows of the towering giants is the shrub layer, with plants spanning from 5 to 20 feet in height. In the sparse light of the forest floor are another set of plants adapted to the dim, still environment, seldom reaching over 5 feet. Linking these layers together and providing roadways for the wildlife are herbaceous vines and woody lianas that form a twisting, interconnecting web. Epiphytes, living on branches in all the strata, are salient components of the tropical forest. Gaps generated by storms and disease allow the genesis of secondary growth, which for a period contribute their own special habitat and life forms before eventually becoming part of the primary forest. It is assumed that the amber woods had these basic features as well.

Tropical forests today cover only 7 percent of the land surface, yet contain about 50 percent of all living species.[11] How long these woodlands have prevailed is unknown, but it is clear that the flora and fauna that characterizes each one is indirectly dependent for its survival on soil and climatic conditions, especially moisture. An examination of the vegetation is the surest way of defining the forest type, and the plants determine what types of animal life can survive in that ecosystem. Beginning our journey into the past, we will now examine plant and animal clues in amber to elucidate the mysteries of the forest that was the home of our bee.

2

The Amber Forest

The Plants

The most significant plant in the ancient forest was the extinct species of algarrobo tree that was the source of Dominican amber.[1] Without this tree, all of the life forms described here would have been lost. Aside from trapping and preserving organisms in its resin, the tree provided leaves, pollen and fruits for food, support for a range of epiphytes, and homes for diverse animals, from bees to mammals. Algarrobo petals, stamens, leaves, developing fruits, and buds drifted into the resin on the bark. This tan-colored petal (1) probably fell soon after the flower opened. Perhaps a bee flew against it while collecting pollen or maybe it simply detached because the flower was fertilized and its short life was finished. Once free, the petals, one after the other, drifted into the still, humid air, brushing against leaves and bark. Sometimes they would be caught on spider webs or on a mass of sticky resin on the trunk or just settle to the ground, carpeting the forest floor. The petal is so well preserved that small brown resin pockets scattered throughout the tissues are still visible. The delicate mechanisms of amber preservation are illustrated in this picture of a stamen (2). As it struck the surface of the resin, a shower of pollen grains were released from the anther. These grains are still intact, containing the original protoplasm. The algarrobo fruit was a pod similar to those of some locust trees today. The pod shown here (3) represents the most mature stage found. Still relatively young, the seeds were probably undeveloped, whereas the lumpy, outer

1. A petal from the algarrobo flower belonging to the extinct species *Hymenaea protera* (Leguminoseae). The small dark resin glands stand out against the light-brown tissues (C).

2. Pollen grains cascading out of a stamen from the algarrobo flower when it struck the fresh resin. Detailed examinations of pollen in amber still await investigation (C).

3. A developing seed pod from the algarrobo flower. This pod is still attached to the flower stalk and surrounded by calyx lobes (C).

4. A developing flower bud from the algarrobo tree. This appears fresh enough to open if placed in water.

surface had already acquired the characteristics of the ripe fruit. This bud (4) contains an unopened flower. Locked within it should be three large petals, two reduced scale-like petals, ten stamens, an ovary with a tiny pod, and four sepals.

The natural history of the extinct tree was probably similar to that of its relatives now living in Central and South America and Africa. Present-day algarrobo trees are confined to the tropics but exist in dry as well as wet forests.[2] These trees can grow to heights exceeding 120 feet and sometimes emerge above the canopy layer. Each tree flowers over a period of 4–6 weeks, usually at the end of the dry season, with only a small number of flowers opening at a time. Flowers are pollinated by bats as well as bees, wasps, moths, and butterflies, and the petals and stamens fall soon after pollination, which was probably also the case with the extinct algarrobo. The fruits develop into hard, brown, indehissent pods which eventually fall to the ground. Seed dispersal is assisted by animals such as agoutis and peccaries that break open and eat the pods.

The algarrobo probably was a dominant canopy tree in this prehistoric realm, and although a few scattered algarrobo trees exist in the Dominican Republic, they are quite different from those of the past. All produce resin from glands that occur in their petals, leaves, fruits, branches, and trunks, the purpose of which has been the subject of much speculation. It has been demonstrated that the resin is repellent to some insects and fungi and the algarrobo may have its own built-in pesticide.[3] To understand its usefulness, one must examine the conditions under which this group of plants evolved. If they did evolve in the late Cretaceous, perhaps the resin was a deterrent to grazing dinosaurs. Vertebrate herbivores today certainly do not seem to relish the mature leaves of algarrobo trees.

Thrusting up into the canopy, competing for sunlight with the algarrobo, grew the cativo tree (5), whose modern descendants grow to 120 feet.[4] Because of the proximity of these two trees, it is not unusual to find the flowers of this legume captured in amber. The small flowers of the cativo actually lacked

A halictid bee visiting flowers of the algarrobo tree. Depicted are petals, stamens, developing ovaries, and a bud of this extinct legume (see photos 1–4).

5. A pair of flowers of the cativo tree belonging to the genus *Prioria* (Leguminoseae) (C).

6. A flower from the nazareno tree, a member of the genus *Peltogyne* (Leguminoseae).

petals, but the five sepals formed a stellate cup enclosing ten fragile stamens and a hirsute pistil. Cativos have clusters of flowers scattered along the tips of the branches. During the dry season, these flowers clothe the tree in bouquets of tiny milk-white floral sprays. The coarse bark of cativo trees exudes a sweet black gum, whose aroma hangs on the humid air.

These trees, although dominant canopy trees long ago, are now absent from the Dominican Republic but still occur in Costa Rica, growing in lower montane and evergreen lowland forests and in Panama and Colombia in tropical wet and moist forests. White-faced capuchin monkeys harvest the fruits in the canopy while peccaries forage for the flat, coffee-colored dehiscent pods in the forest litter.[5] The buoyant seeds are found floating along the edges of dark shadowy ponds and drift along the shores of lakes, where they await the opportunity to resume life again.

Another tall legume whose slender branches mingled with those of the algarrobo and cativo was the nazareno (6). The tiny flowers with five delicately veined petals cradled ten out-stretched stamens waiting to make contact with any small insect that came to sample the nectar. Some insects, such as tiny thrips, stopped to consume the pollen while defoliators nibbled on the leaf edges, leaving damaged leaves in amber as evidence of their passage. Descendants of the nazareno occur in tropical moist and wet forests in Central and South America but are gone from the entire West Indies.[6] These trees reach heights of 150 feet, and their buttressed bases can extend up to four feet in diameter. Perhaps the wood of the primeval nazareno was a rich magenta like that of its descendants. During the short blooming period, the fragrance of the tiny snow-white flowers permeates the forest, and later the compressed, ocher seed pods contrast sharply with the light green foliage of the new-generation leaves.

From the frequency of their floral remains, it can be concluded that the algarrobo, cativo, and nazareno comprised the dominant trees in the canopy of this early landscape. They probably grew side by side, their branches almost touching, forming a nearly unbroken foliage rooftop throughout the forest. Evidence from amber tells us that there were other canopy trees present, but since relatively few flowers have been found, it is assumed that these trees were less abundant than those mentioned above.

One relatively scarce tree, the sebo, had extremely small flowers. This sebo flower (7), a male formed by aborting the female portion, possesses three folded stamens. Extant sebo trees attain heights of 120 feet and are characterized by flat-topped crowns. Their horizontal branches bear clusters of minuscule golden flowers far over the forest floor. The leaves, however, make up for the tiny flowers by reaching a foot in length. How the fragile female flowers are pollinated is a mystery; possibly this is accomplished by small insects or by a breeze that wafts through the canopy. Small nutmeg-like fruits bear fat-laden seeds that are relished by monkeys and birds that are probably instrumental in distributing these trees. The sebo is gone from the

8

7

7. The sebo tree had small unisexual flowers. Here a male flower aborts the ovary. A member of the genus *Virola* (Myristinaceae), it is related to the nutmeg.

8. A flower of the caoba tree belonging to the genus *Swietenia* (Meliaceae), a close relative of the mahogany tree.

Greater Antilles, but present populations occur in wet, moist, and premontane rain forests in Central and South America.[7]

Another tree that could have poked through the canopy as an emergent is the caoba (8). Five small petals encircle the stamens of the ancient flower. The white to red-tinted blooms of modern species open at dusk, and a variety of moths serve as pollinators when they come to sample the nectar. Orange to scarlet seeds that survive the ravages of weevils are relished by birds and mammals, who then distribute the plants. The leaves are notched by the larvae of colorful swallowtail butterflies. The coaba tree is closely related to the mahogany, and its foliage may have given a diffuse dark green hue to the amber forest, similar to those trees in Hispaniola today that have survived the drastic over-cutting.[8] But unfortunately it is no longer possible to see the old mahogany giants of the past that towered to heights of 135 feet and reached 6-foot diameters.

Among the subcanopy trees of the early woods were the souca and sigua. The five petals of the souca flower (9) were

A flattened palm bug (photo 22), belonging to a group known today that lives only on palms, is on a leaf of its likely royal palm host plant depicted in the background.

9

10

9. A compact flower of the souca tree, a member of the genus *Trichilia* (Meliaceae) (C).

10. A flower of a sigua tree belonging to the genus *Nectandra* (Lauraceae).

small and contained ten stubby stamens, similar to those of their descendants in the Dominican Republic today.[9] The rare sigua (10) had petite, six-part flowers and a characteristic, complicated arrangement of stamens and glands. Modern sigua trees grow to heights of 80 feet, and their greenish white flowers mature into dark cherry-sized fruits that nourish a variety of animals. They persist in tropical moist and premontane forests with several species remaining in Hispaniola.[10]

Other plants that probably reached into the subcanopy realm were palms and fig trees. Flowers of palms as well as palm-specific bugs (discussed later) occur in amber and provide both direct and indirect proof of these majestic trees in the ancient forest. The white, stout, solitary trunks of the palms must have stood out against the background greenery. From the graceful crown draped feathery fronds, providing unique perches for the forest animals. Stingless bees swarmed around the flower clusters, and it is likely that bats distributed the fruits, similar to the scenario with present-day royal palms.[11] The presence of figs in the primeval forest is provided by indirect evidence of their pollinators (discussed later), and it is likely that some of these figs were large-sized trees that at least reached the subcanopy layers. Characterized by shiny leaves and far-reaching limbs, these plants also provided an

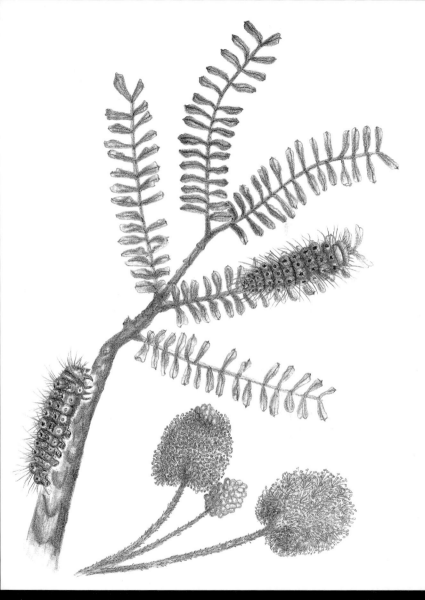

Flowers (photo 11) and leaves (12) of an acacia tree, many types of which occurred in the amber forest, showing leaf-feeding riodinid butterfly caterpillars. These riodinids belong to groups that did not have associations with ants.

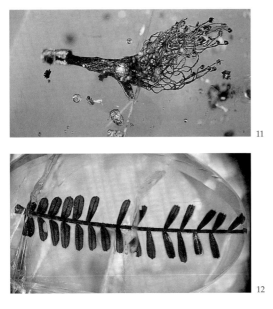

11

12

11. Acacia flowers (Leguminoseae) such as this one must have added color and grace to the amber forest (C).

12. This pinnatifid leaf came from one of the mimosoids (Leguminoseae). Many types, especially representatives of the genera *Acacia* and *Mimosa,* occurred in the amber forest (C).

important source of nutrients in their fleshy fruits to the forest animals.

The understory layer of this venerable tract contained plants adapted for survival in the shadows of the canopy giants, which competed for the small patches of sunlight that filtered through the branches and lianas above. Among those claiming the understory as their domain were the mimosoids. Today these plants (acacias, mimosas, and their relatives) have developed a wide diversity of growth habits: small herbs hug the forest floor, short to tall bushes and trees explore the spaces below the canopy, and vines find sunlight by climbing up and over the limbs of other trees.[12] Although their habits differ greatly, they are characterized by their roundish clusters of stamen-dominated ivory, saffron, or rosy-pink flowers (11). Many species left leaves (12), leaflets, and small flowers in the resin.

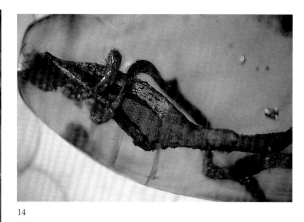

14

13

13. A flower belonging to a member of the family Thymeliaceae (C).

14. Twisting tendrils that appear as gnarled serpents are evidence of vines (C).

Modern mimosoids grow in the understory of dry to wet tropical forests. Unfortunately, from their remains, it is not possible to determine whether the plants were trees, shrubs, herbs, or climbers, but certainly many of them inhabited the understory layers.

A member of the family Thymeliaceae (13) has been identified through its beautiful rare flower featuring a long calyx tube. The pistil extends far beyond the stamens, perhaps to avoid self-pollination. Members of this family are mostly understory trees.

Lianas and vines (14) must have been a dominant feature of this pristine kingdom, their thick stems dangling from the canopy heights. Some vines probably trailed straight down while others wound their tortuous way over and around, using various plants for support and themselves serving as pathways for insects, scorpions, lizards, and small mammals. Today, some lianas begin their life in the soil, groping steadily upward toward the light of the canopy; others germinate in small fertile niches on the canopy trees and send their stems earthward. During the blooming period, tiny flowers dot portions of these

15. This flower belongs to a member of the genus *Peritassa* of the family Hippocrateaceae (C).

woody vines, displaying flashes of milky-white and vivid yellow along their dark serpentine stems. The liana flower shown here (15) belongs to a genus known as *Peritassa*. Like its descendants in South America that occur in moist and wet forests,[13] this ancient plant probably had a woody stem and sinuous, tendril-like outgrowths that twisted around the branches of the algarrobo tree. Opposite-toothed, jagged leaves contrasted sharply with the smooth-edged leaves of the canopy trees. Its diminutive flowers, with five perfectly formed sepals and petals, undoubtedly attracted a host of insects that came to imbibe the nectar. When ripe, the capsular fruits characteristic of modern *Peritassa* species open to free winged seeds that slowly spiral down to the forest floor, hopefully landing in a spot favorable for germination.

In the shrub layer of this bygone realm were probably dwarf palms that may have reached 20 feet in height. Many of these were likely to be fan palms which provided perches for birds and lizards as well as dry undersurfaces for wasps to construct their paper nests.[14]

On the amber forest floor grew pega-pega, a broad-leafed herbaceous bamboo. Its presence would never have been detected if it hadn't been for the unusual manner by which this grass is distributed. Small hooked hairs on the spikelet tips of extant pega-pega (16) stick to the pelts of animals and probably to the feathers of birds. In the present case, a mammal picked up some of the spikelets as it was wandering through the

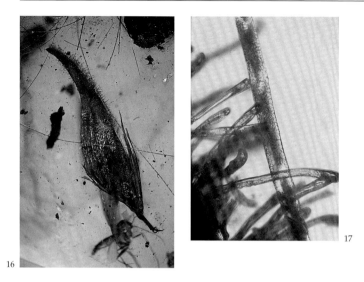

16. A pega-pega bamboo spikelet of the genus *Pharus* of the family Gramineae (C).

17. This pega-pega spikelet contains hooks for clinging to the pelage of birds and mammals. Hairs are still attached to the spikelet and provide clues to how this flower became embedded in resin (C).

forest. Finding them irritating, it rubbed against the trunk of an algarrobo tree in an attempt to brush them off. At least one came off, thanks to some sticky resin on the bark. Some of the animal's hairs came off with the spikelet (17), enabling us to piece together this unusual story and identify the carrier as a possible carnivore.[15] Spikelets of present-day pega-pega have been recovered from the pelts of jaguars in Ecuador.[16] This bamboo still occurs in the Dominican Republic today, as well as in other tropical moist and premontan wet forests in Central and South America.

Other clues to the flora of this primeval region are plant products such as pollen, fruits, and seeds. Studies on pollen are still in their preliminary stages because practical methods to free the embedded pollen grains from the amber matrix have not yet been developed. Fruits and seeds do occasionally appear but they are usually small and difficult to identify un-

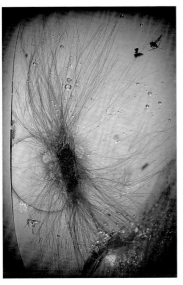

18. Seeds like this one, which came from a member of the family Bignoniaceae, provide important clues to the amber forest flora.

19. The source of this unusual seed in unknown (C).

less one is familiar with that particular plant species. These two rare seeds have interesting appendages for dispersal. One is elongate, just under 10 mm in length, including the hairlike ends (18). There are several hundred elongate hairs on both ends of this particular seed, making it resemble a patch of furry skin. Probably belonging to the family Bignoniaceae, these plants are common epiphytes in tropical forests.[17] A strikingly beautiful "spiked ribbon" seed (19) has elongate golden ribbons radiating out from a coiled central hub and measures 15 mm from end to end. The projections on these seeds are obviously important for dispersal, but exactly how they function is a mystery. An infinitesimal seed has been identified as possibly belonging to an orchid (20). The seed is barely over one millimeter in length and has fragile "wings" on both ends. Orchid seeds are minute and primarily dispersed by wind, so they could have readily been blown against some sticky resin.

This jaguar-like feline is nestled among a clump of pega-pega bamboo plants (photo 16). In depicting this carnivore, some artistic license has been taken since we cannot be certain of the exact identity of the carnivore hair in the amber, and whether carnivores roamed the West Indies in ancient times is controversial.

20. Even minuscule seeds (only 1.3 mm long) like this one, which could belong to an orchid, can establish the presence of a family of plants (C).

21. The source of this pair of fruits is unknown, but they could belong to an epiphyte (C).

Fruits are even rarer than seeds. An amazingly well-preserved pair of berrylike fruits (21), each supported by a cup-shaped calyx, captivates the eye. The plant that produced these may have been an epiphyte that clung to the bark of the algarrobo. After blooming, the tiny fruits were possibly eaten by birds who then distributed the seeds to other locations. It's easy to imagine that as a bird hurriedly stripped the fruits from the plant, some escaped its beak and landed in the sticky resin.

Any plant remains, no matter how inconspicuous and seemingly insignificant, can be used to determine the wide diversity of vegetation present long ago. Plant hairs or trichomes are microscopic outgrowths from the epidermal layer of buds, leaves, and young twigs. They occur in a wide variety of sizes and shapes, and different plant families produce their own types. Some are flat with scalloped-shaped edges, others may be star-shaped, and many are simple hairs. Some T-shaped trichomes were seen attached to the underpart of a planthopper in amber. These particular trichomes are characteristic today of plants in the Malpighiaceae, a family that includes small to medium-sized trees and lianas.[18] From this microscopic evidence, a new family of plants is added to the list of the ancient

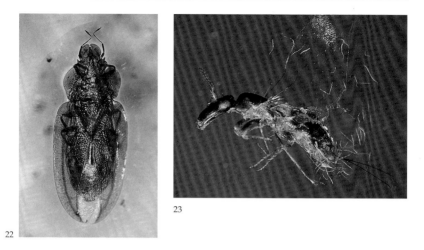

22. This extinct species of palm bug, *Paleodoris lattini* of the family Thaumastocoridae (Hemiptera), provides indirect evidence of palm trees in the amber forest. Its smooth, flattened body enables the insect to live between the sheaths of the unopened palm leaves (C).

23. A female fig wasp belonging to the family Agaonidae (Hymenoptera) provides definitive proof of *Ficus* (Moraceae) in the amber forest. Note the nematodes (*Parasitodiplogaster:* Diplogasteridae: Rhabditida) that were concealed in her body cavity, hoping to reach the safe refuge of a fig inflorescence (C).

flora, and something about the habits of that particular planthopper becomes known.

Indirect evidence of the existence of plants can also be obtained from finding insects that are dependent on specific types of flowers, fruits, or leaves. One example is the curiously flattened palm bug (22) that possesses the same features as its modern-day descendants that live between the leaflets of royal palms.[19] The compressed body of these bugs allow them to squeeze between the tightly closed leaves of young palm fronds. Even the legs and rostrum are flattened. In this secluded and protected habitat, the palm bugs feed, mate, deposit their eggs, and die.

The presence of small, delicate fig wasps (23) is yet another example of how insects can inform us about the existence of certain plant groups in a bygone forest. Members of the fig genus *Ficus* include a number of shrubs and trees.[20] Each species has its own specific wasp pollinator, and, in return for pollination, the insect is provided with a secure place to rear her young.[21] Shown here is a female, and, on the basis of what we know about this association today, we can say she was searching for the right species of fig to enter. Somewhere on her body (often in pockets or special crevices) are stored pollen grains from the fig in which she was born, and she will use those grains to fertilize the plant in which she hopes to deposit her eggs. To help her achieve this goal, receptive flowers that are ready to be pollinated emit a chemical odor that their bene-factors recognize. After locating the fig, the intrepid creature begins the difficult task of entering the tightly closed chamber via a pore that leads through a tunnel of overlapping bracts. Using mouthparts, thick front legs, and even antennae, she forces her head between the bracts and slowly struggles down toward the open flower cavity. During this process, wings and antennae are often torn from her body. Such abuse can be over-whelming and she may die, trapped in the maze of bracts. Those that do run the gauntlet and reach the dark hollow re-ceptacle find their reward, a host of tiny receptive flowers lin-ing the walls of the chamber, some of which have short styles (top portion of the female part of the flower) and others long. During the oviposition process, the female inserts a needlelike ovipositor through the top and lays an egg in the developing seed. As the egg passes from her body, it flattens into a long, flexible, cylinder that allows it to flow through the ovipositor. The female's ovipositor can reach only the short-styled flow-ers, thus the flowers containing wasp eggs develop into insect galls, while the long-styled flowers, produce normal seeds. During the process of oviposition, the female releases pollen from her body that will fertilize the flowers, and she dies soon afterwards while still inside the receptacle. Without such wasps, fig flowers would not be pollinated. Within the insect flower galls, the eggs hatch into larvae that feed on the plant tissue. After the larvae have completed their development, they pupate and emerge as adults. The males emerge first. Their purpose is twofold, to fertilize the females and to assist them in escaping from the confines of their growth chamber. Wingless, light-colored, and with reduced eyes and antennae, they never leave the flower cavity. Their strong mandibles play a vital role in carrying out their mission. They have no trouble finding a gall containing a female, and with their strong jaws they gnaw a hole in the side, insert their abdomen, and mate. The fertilized females wait patiently in their cells while the males complete their ritual and make a hole in the wall of the fruit. Before their escape, the females walk over the newly emerged stamens of the florets, collecting pollen to take to the next plant where they will deposit their eggs and pollinate the flowers. Each species has its own unique way of carrying this pollen, ensuring that it is not lost along the way.[22]

Even strangler figs, which begin their life as epiphytes on the canopy trees wherever frugivores have dropped their seeds, need wasps for pollination. The small innocent-looking seedling sends down a root which can reach the astonishing length of 120 feet before making contact with the forest floor. Once properly rooted in the ground, the plant expands, sending out shoots that eventually form an enclosing network around its host tree. Finally, this monstrous plant, with its humble beginnings, destroys the supporting tree and stands alone, not having had to compete with the other canopy plants to reach the top. Such are the benefits of the strange symbiotic association between this towering plant and its 2-millimeter-long pollinator.

Through the millennia, other symbiotic associations have developed in relation to fig wasps. Some creatures discovered that the receptacle contains a rich source of nutrients, but there was no way of reaching that supply except via the wasps which offer a perfect opportunity to reach the interior. A small group of nematodes (tiny wormlike creatures) have found ways to conceal themselves under the overlapping segments on the pollinator's body. In this way the insects carry the nematodes

Specific pollinators of figs, a female fig wasp (photo 23) approaches a fig receptacle where her young will be raised. Her sharp mandibles would have been used to assist her in entering the fig, and her antennae and wings would probably have been lost in the process. Note the fine membranous wings, and the needlelike ovipositor used for inserting eggs into the fig florets.

from receptacle to receptacle. Once inside the flower cavity, the nematodes exit their ride and multiply. Their life cycles are synchronized so that when the female wasps emerge, the phoretic stage of the nematode is there, waiting to crawl under her segments and be carried to another plant. In Panama today, all species of fig wasps so far investigated have their own type of affiliated nematode belonging to the genus *Parasitodiplogaster*.[23] That this is an ancient alliance is shown by the presence of such nematodes emerging from the fig wasp depicted here.

An insect that provides indirect evidence of epiphytic bromeliads is a stalk-winged damselfly.[24] Its descendants now lay their eggs in water collected inside leaf reservoirs of tank bromeliads. These quite unusual plants have solved the problem of desiccation by collecting rainwater in cavities formed at their collective leaf bases. An amazingly diverse assemblage of animals and plants can survive in these pools, which sometimes contain only small amounts of water. Single-celled protozoa and algae abound in these microcosms, along with tadpoles and a host of other invertebrates. Voracious inhabitants include the larvae of damselflies, which attack and devour many of the other occupants. The long abdomen of this damselfly could be an adaptation for ovipositing in deep, narrow recesses. It is possible that she was searching for a bromeliad in which to deposit her eggs when she accidentally brushed up against some resin. More discussion about tank bromeliads appears in the section on aquatic insects.

Additional evidence of epiphytic bromeliads and orchids is presented by butterflies and orchid bees. The butterflies (53) belong to a genus in the metalmark family whose descendants today feed only on epiphytic bromeliads and orchids.[25] Their caterpillars are covered with a dense protective layer of long setae. Orchid bees (121) imply the presence of orchids in this lost world. While female orchid bees collect pollen and nectar from a variety of plants, male orchid bees collect aromatic secretions from orchids and in so doing pollinate them in very specialized ways, so much so that they are the only pollinators of many orchids.[26]

While this female orchid bee (photos 121, 122) may visit orchids for nectar, it is the males who function as specific pollinators for many of these plants. These bees have greatly expanded hind tarsi, which are used by the females in transporting pollen and by the males in storing fragrances removed from orchid flowers.

Epiphytes were a notable part of the former biota. The branches and trunks of the canopy and understory trees offered a wide variety of microhabitats for plants. Not only were bromeliads and orchids covering the exposed bark of the trees, the dominant diminutive cover (known as epiphylls) was composed of cryptogams, small plants that lack true flowers and reproduce with spores. A dense carpet of cryptogams covering a few inches of bark would have included ferns, liverworts, mosses, lichens, algae, and fungi. Representatives of all of these plant groups nestled together on the branches and trunks, growing over and sometimes parasitizing each other to survive. Some of the ferns, like aromatic grammatids (a genus of small, epiphytic ferns) with their spore-bearing fronds reaching 20 cm, could survive well up on the limbs of the canopy trees. The spore masses of grammatids were quite conspicuous on the backs of their leafy green pinnae, giving the plants a speckled appearance. But like the mosses and liverworts, these ferns could also reproduce vegetatively by forming little buds on the tips of the leaves, which would drop off and start another plant.

Quite an assortment of bryophytes dominated this ancient arboretum, representing the largest fossil bryophyte flora recorded in the past 65 million years.[27] Mosses and liverworts, which probably nestled around the base of fern fronds, produce a characteristic minty scent imparting to the forest understory a unique, pungent odor. All of the bryophytes found thus far are typical of tropical forests. Some preserved mosses and liverworts (24) bear tiny reproductive buds called gemmae

24. A liverwort belonging to the genus *Bryopteris* (Lejeuneaceae).

Portion of a tree trunk with a newly opened mushroom (photo 29), accompanied at its base by a liverwort (24) with gemmae on the tips of its leaves (25), and a "dead man's fingers" fungus in the lower left (31). A small mite (30) on the stipe of the mushroom is flanked by two mite predators, an antlike stone beetle (67) in the lower right, and a short-winged mold beetle (68).

25. On the edges of the *Bryopteris* leaves are small reproductive structures, representing the first fossil record of gemmae on a liverwort (C).

26. A moss of the genus *Hypnum* (Bryophyta). Still enclosed among the leaves is a small mite (Acari).

(25), which eventually fall off and initiate new plants. Such adaptations today are typically formed during arid periods, thus providing evidence of dry seasons in the amber forest.

Within mats of moss (26) were concealed inhabitants of the original epiphyll community, including plant-parasitic nematodes which crawled over the leaves, stopping here and there to puncture the walls with their microscopic syringe-like mouths and then suck up the cell juices. Other types of nematodes ingested the bacterial and alga cells growing on the moss surfaces, vacuuming them up in tubular mouths with rhythmic contractions of their muscular pharynx. Competing with these nematodes for food in this habitat were rotifers that ingested microorganisms.[28] Dwelling in films of water covering the leaves were amoebae that sent sprawling pseudopodia in all directions, entrapping microscopic life. These in turn were hunted by capacious sprawling slime molds that periodically patrolled the area, engulfing everything in their path. The

Growing on this log are portions of a lichen (photo 27) complete with isidia and thallus, and a moss (26) with gemmae on the tips of some of its leaves. Lurking in the vicinity is an oribatid mite (66), while a polyxenous millipede (28) is heading toward its food source.

27

28

27. A lichen fragment belonging to the family Parmeliaceae (Ascomycetes). Small protrusions called "isidia" dot the surface of the thallus (C).

28. This adult millipede of the family Polyxenidae (Polyxenida) is one of the smallest members of the Diplopoda, measuring barely 1 mm in length. It is well camouflaged while resting on lichens (C).

giants of this world, mites (26) and springtails, browsed over the plants, feeding on spores, nematodes, and other life forms.

In addition to liverworts and mosses on bark and leaf surfaces were lichens (27). Because most lichens are comprised of a symbiotic association between a fungus and algae, they also are able to tolerate exposed habitats on rock and wood surfaces inhospitable to many other plants. There to browse on the lichens were miniature hairy millipedes (28) that looked like tiny beetles with thirteen pairs of legs. These millipedes, scarcely a millimeter long in the adult stage, were well camouflaged, and when danger threatened they could flatten themselves against the substrate, appearing like just another bump on the bark.

Here and there among the lichens and mosses flourished tiny mushrooms and strange branching clusters of fungi known as dead man's fingers. The fruiting structures of these fungi probably never lasted more than a day or two, just long enough for them to release their spores. This tiny mushroom (29) is one of the smallest members of the inky cap family.[29] It probably was clustered together with others of its kind on the bark of the

29. This mushroom is the extinct *Coprinites dominicana* of the family Coprinaceae (Basidiomycetes). It is the only known fossil tropical mushroom (C).

30. This mite, a nymph of the genus *Teleioliodes* (Oribatida), was on the cap of the mushroom when a resin flow covered both organisms.

algarrobo tree. But it did not escape notice there. On the cinnamon-colored cap grazed a tiny mite (30). When the resin flow started to cover its meal, the mite held tenaciously onto the edge of the cap with its front legs and now remains in nearly that same position. Other life forms associated with this tiny mushroom show the complexity of the bark community in the far past. Foraging between the gills was a group of microscopic rotifers, probably dwelling in the moisture film clinging to the mushroom and entering the gill spaces in search of single-celled prey. Even these organisms did not live serenely, since lurking in their midst was an unidentified predator equipped with movable spines around its mouth, poised and ready to impale any unsuspecting rotifer.[30]

The structure known as dead man's fingers (31) is actually an ascomycete fungus that is fairly common in tropical forests today. The spores germinate on many types of substrates composed of decaying vegetative matter. The different colors of the fungal body indicate the two types of spores, sexual and

31. This "dead man's fingers" fungus belongs to the genus *Xylaria* (Xylariadeae: Sphaeriales: Ascomycetes). The white areas produce asexual spores and the dark areas sexual spores (C).

asexual, that are produced. Undoubtedly, these and most other fungi present were utilized as food by both invertebrates and vertebrates.

As noted previously, it is feasible to detect the presence of different plants in the amber forest from direct and indirect evidence obtained from fossils. But a much greater floral diversity of plants existed in the ancient forest than can be presented here. Space allows us to include only the plants considered important for this reconstruction, necessitating the omission of many additional families. To this list we can add plants that have not yet been recovered in amber as well as those whose habits would make them very unlikely to end up fossilized.

In this early silva, the canopy formed its own world of flowers, fruits, and leaves, together with all the animal life that depended upon it for sustenance. Practically every day, especially during the wet season, there were periods of rain, often accompanied by claps and peals of thunder, blinding flashes of lightning, and gusts of wind that tore flowers and fruits from the trees and lianas. During these intervals, the constant din of insect life would cease while most took shelter. After the storm passed, the insects would resume their feeding activities. Although it is difficult to obtain direct fossil evidence of insect herbivory, signs of feeding activity have been preserved. Some

33

32

32. A skeletonized leaf shows that insects had recently fed on this plant.

33. A leaf beetle in the genus *Walterianella* (Alticinae: Chrysomelidae: Coleoptera) might have been responsible for the feeding damage shown in photo 32 (C).

insects skeletonize leaves, eating away the tender parts and leaving the tougher veins untouched (32). One type of insect with this behavior is the flea beetle (33), which belongs to the largest group of leaf beetles. It is estimated that there are between 8,000 and 10,000 species living today. All have a special way of avoiding predators—by jumping. In fact, because they possess a special springing device in their enlarged hind femurs, they are able to leap great heights with lightning speed.[31] As a group, the flea beetles rival the obnoxious fleas as the best jumpers in the insect world. Such acrobatic abilities enable the beetles to avoid enemies that might detect them feeding on the leaves of various trees and lianas. An evasive vault may take them to another leaf, down to the forest floor, or end up in flight.

Scallop-shaped notches made on the sides of a leaf (34) represent another example of insect damage. Soon after an insect feeds, the leaf seals the wound to prevent the entrance of fungi

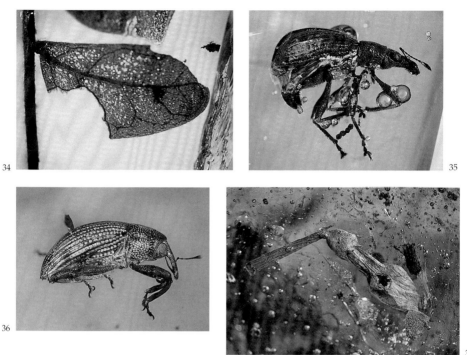

34. A partially eaten leaf had responded to the damage done by an insect (C).

35. A weevil of the subfamily Polydrusinae (Curculionidae: Coleoptera) fed on leaves (C).

36. A number of weevils of the family Curculionidae (Coleoptera) lived in the amber forest (C).

37. A hole in the side of this developing flower bud was made by an insect in search of pollen (C).

and other organisms. A likely candidate for this feeding damage could be a weevil (35) that avoided predators by coming out at night to feed. Mating and oviposition would have been her other activities. Many other types of weevils wandered through this ancient terrain (36).

Flowers were an important source of nourishment and signs of insect activity appear on algarrobo petals, sepals, stamens, and developing pods. Even the delicate nips of a snail have been identified on one petal. Insect damage to an unopened flower bud (37) is shown here. The insect was intent on boring

Leaf beetles (photo 33) skeletonize a leaf (32), while a polydrusine weevil (35) notches another (34), and a zygopine weevil (38) that has made a hole in a flower bud (37) feeds on the tender floral parts inside.

38. With eyes near the top of its head, this zygopine weevil (Curculionidae: Coleoptera) has a snout well fitted to enter plant tissue (C).

a hole through the side to reach the tender floral parts. A reasonable contender might have been one of several zygopine weevils (38), which, with their mouthparts located on the tip of their snouts, have a built-in detection device for predators. Their eyes are positioned near the top of their head, so when they detect the slightest movement from above they can plummet like a dead weight straight down and remain out of sight on the forest floor.

Plant-Feeding Insects

An abundant group of plant-feeding insects in the amber forest were the Homoptera. In temperate climates, one is familiar with aphids, the annoying pests on ornamental plants. However, in tropical forests, aphids are uncommon for some obscure reason. Their niche is occupied by a host of others such as planthoppers, leafhoppers, and treehoppers. Planthoppers come in a variety of shapes and sizes and represent some of the most elegant as well as bizarre insects.[32] Those from long ago were no exception. Their constantly flashing, brightly colored wings must have glittered in the canopy layer. They added dissonance to the myriad of insect sounds filling their primitive domain with rasping and chirping noises interspersed with staccato-like cadences complete with diminuendos and crescendos. Possessing sucking mouthparts enabled them to use their needle-like beaks to penetrate plant tissues and draw up the juices. While the leaf remains intact, discolored areas provided evidence of their visits. Extant planthoppers prefer

39

40

39. An amazing variety of planthoppers occurred in the amber forest. This representative of the family Delphacidae (Homoptera) did not have time to retract its wings (C).

40. This ornamented issid planthopper (Issidae: Homoptera) does not have fully developed wings (C).

the tender tips of new leaves or shoots, and in a tropical forest, where many trees shed their leaves at least once and often twice a year, there is a continuous supply of recent plant growth to feed upon. When new growth is plentiful, some planthoppers produce wingless forms which remain on the plants. When that food source is exhausted, winged forms appear and then disperse to other trees. Plant damage is not just from feeding, however, since many female planthoppers have a saberlike ovipositor used to pierce plant parts in order to deposit elongate eggs in the wound.

Some planthoppers, the delphacids, have antennae that protrude from their heads (39). Other planthoppers like the issids are known for the mulberry ornamentation that occurs on the sides of their heads (40). A few derbid planthoppers have their wings expanded at the tips (41). There are issids that are built quite sturdily (42) and other fulgorid planthoppers have elongated heads (43), so much so that is difficult to tell what's front and back. These are called "alligator-headed" or "dragon" insects, since the snout resembles to varying degrees the head of

41. A planthopper of the family Derbidae (Homoptera) shows a curious expansion at its wing tips (C).

42. An issid planthopper (Issidae: Homoptera) shows how compact some of these insects can be (C).

43. The head of a dragon insect (Fulgoridae: Homoptera). Do such forms really frighten away predators? (C).

an alligator. Naturalists have noted that modern types of such planthoppers often sit with their snouts up in the air, similar to the stance of a true reptile.[33] Whether this behavior actually frightens potential predators is unknown, but why else would such a posture evolve? Another tactic used by planthoppers to escape predators is patterned camouflage designed with colors that blend into the foliage. Some forms have wing veins that mimic plant leaves.

Many insects have the ability to produce wax in various forms. Some issids developed a novel way of escaping from

Planthoppers were abundant in the ancient forest, and most expired almost immediately after landing in the resin, some with wings extended (photo 39); others had undeveloped wings (40). Most spectacular were the dragon insects (top left) with their elongated heads (42).

44

45

44. A shaving-brush planthopper (Issidae: Homoptera), with a tail full of wax filaments to confuse its enemies (C).

45. A leafhopper of the genus *Xestocephalus* (Cicadellidae: Homoptera), one of the most common leafhoppers in Dominican amber (C).

their enemies and might be called "shaving brush" plant-hoppers (44). The young produce a "tail" of long wax filaments from an area near the tip of the abdomen. These strands can protrude a distance equal to, or even greater than, the length of the insect. Being refractive, they are very noticeable, causing a predator to strike at the tail. The predator ends up with a mouthful of waxy filaments while the now "tail-less" plan-thopper darts away. The fossil tail-less survivors that have been found suggest that at the moment of attack, the leap for safety took them toward a pool of resin.

Leafhoppers abounded in the upper strata of the vanished forest. They were smaller than most planthoppers and their spiny hind legs were a characteristic feature. They fed on the juices of the canopy plants. The leafhopper pictured (45) here was minute but probably occurred in large numbers; other leafhoppers were larger and mimicked distasteful or stinging insects. When viewed from the top, this one (46) has dark markings that resemble an ant, an insect avoided by many predators.

Of particular interest because of their unusual protective modifications were the diminutive stout treehoppers (47). The pronotum (anterior portion of the thorax) of many treehoppers has been modified to resemble some structural feature of the

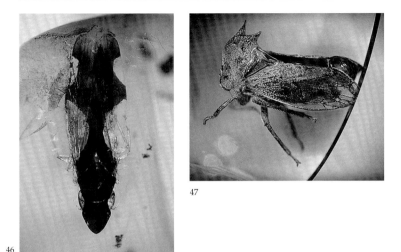

46. The strange pattern of this leafhopper (Cicadellidae: Homoptera) is probably meant to mimic an ant.

47. A treehopper (Membracidae: Homoptera), with its head shaped to resemble a plant thorn, presumably like those that occurred on its plant host (C).

plant upon which they feed. Such modifications often resemble spines or thorns, and these usually match the number and shape of the spines on the plant host. Because these adornments often appear so awkward, these insects are sometimes called "grotesque-looking bugs."

All of the above groups have a trait that can be used to their advantage. They produce deposits of honey dew, an excretory product of their liquid sap diet. This is quite attractive to ants, which relish the deposits enough to protect its source from predators and parasites. Thus an excretory product ended up fortuitously providing a strategy of protection.

Another group of Homoptera haunting that early terrain were the cicadas, well known for the shrill cries of the short-lived adult males during the mating season. The young or nymphs live in the soil for a long time, sometimes up to ten years, feeding on fluids from plant roots. Cicada nymphs are not only slow in developing, they are also sluggish movers; when finished with their development, they emerge from their subterranean

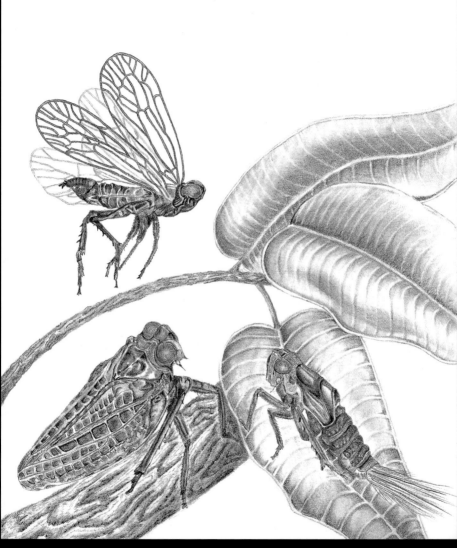

Planthoppers come in a variety of shapes, such as the cixid on the top, with rounded wing tips (photo 41), and the stout issid (42). Many, like the shaving brush issid on the lower right (44), produced wax filaments from the tip of their abdomen as a survival tactic.

48. Evidence of the grotesque is provided by this cicada (Cicadidae: Homoptera) nymph that left its subterranean retreat to transform into a handsome, vocal adult.

49. Monkey grasshoppers (Eumastacidae: Orthoptera), wingless and agile, move through the undergrowth with remarkable ease (C).

home and scale the nearest tree at a snail-like pace. The adults, in contrast, as anyone who has tried to catch one by hand knows, become quite active and can fly off in an instant. The cicada nymph shown here (48) was mired in resin while climbing an algarrobo tree just prior to molting into the adult stage. It has enlarged front legs adapted for digging, imparting a mole-like appearance. One can imagine that at certain periods, especially around dawn and dusk, the primeval forest was filled with the vibrant pulsating screams issuing from choruses of cicadas.

Belonging to an ancient group of plant feeding insects are grasshoppers and katydids, several types of which existed in the primeval forest, although they are quite rare in amber. This one belongs to a group of diminutive grasshoppers (49), commonly called monkey grasshoppers, that frequent treetops but will also rest in the shrub layer in small patches of filtered sunlight. One peculiarity of these grasshoppers is that while resting, they habitually fold their hind legs at right angles to

This leafhopper (photo 45) has its wings fully outstretched. A lacewing bug (55) shows the cuticular patterns that earned the group its name, while a treehopper (47) has three spines near its head that mimics those of the plant upon which it feeds.

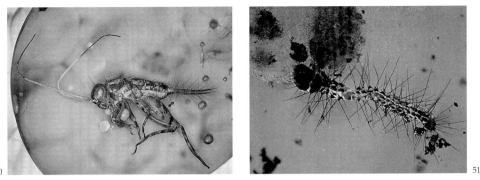

50. Bush crickets, like this male nymph (*Proanaxipha bicolorata*: Trigonidiidae: Grylloptera), were probably quite common in the amber forest (C).

51. This young caterpillar (Lepidoptera) appears well protected with a set of long, stiff hairs (C).

their bodies. This posture gives the impression of two crossing stems, apparently as camouflage against predators.[34] Some instinctively assumed this shape when landing in the resin.

In the understory lived an assortment of crickets, most common of which were the bush crickets (50).[35] Dwelling on leaves and stems of grasses and shrubs, some may have congregated near water sources. Their strong trilling sound, although short, must have added yet another level to the background hum of the silvan concert. The wingless bush crickets spent most of their life on the bark of trees, remaining motionless with their well-camouflaged bodies if enemies approached.[36] Many were nocturnal, although the speckled coloration of some suggest they also ventured out during daylight hours.

Caterpillars (51) occur from time to time. Some are only several millimeters in length and at most can be identified only to family; however, their presence, along with a number of moths, shows that diverse Lepidoptera probably served as pollinators in this foregone realm. One of many groups of moths present were the leaf rollers or tortricids. The caterpillars of these moths made themselves a secure home by curling up leaves and holding them together with silk. Inside they fed on the foliage. Other caterpillars, the tineids, lived in

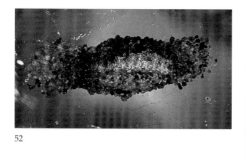

52

53A

52. Some moth larvae (Tineidae: Lepidoptera) prefer to construct protective cases that are carried along as they feed and are used as a refuge when danger threatens (C).

53A. Butterflies such as this metalmark of the genus *Napaea* (Riodinidae: Lepidoptera) are quite rare in amber, but may have been common in the forest (C).

portable, protective cases made out of waste and fecal material (52).

Only a few adult butterflies have been found, and all belong to the metalmark family (53A).[37] Imagine these small, speckled orange-brown butterflies fluttering through the forest canopy and alighting on the vivid fragrant flowers. Some apparently mistook a pool of resin on the algarrobo tree for a morsel of sweet sap. Representatives of this family of butterflies have developed remarkable partnerships with ants, on which the caterpillars depend for protection against predators. These interesting metalmark caterpillars will be discussed further in the section on ants. Unguarded metalmark caterpillars are exposed when feeding and depend on camouflage, mimicry, or chemicals to avoid their enemies. They may be covered with stiff hairs to keep predators away or contain foul-tasting alkaloids or simply look like the leaf of the host plant. Another characteristic caterpillar showed that the metalmarks shared

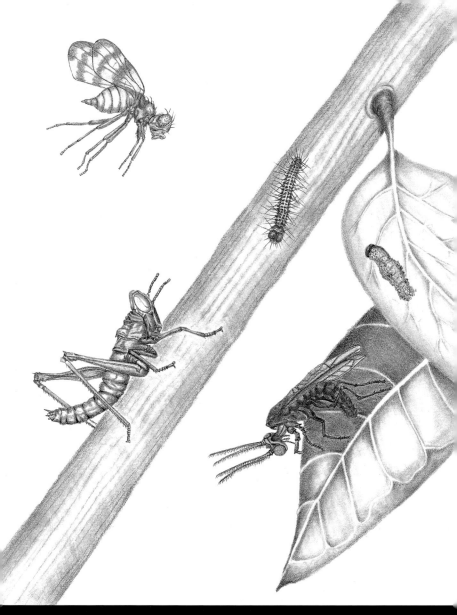

A scene from the understory with a wingless monkey grasshopper (photo 49) resting on a stem. A short distance away is a caterpillar (51) protected by long, stiff hairs. On the adjacent leaf, another caterpillar (52) feeds from inside its protective, portable refuge. On the lower leaf is a male argid sawfly with its antennal branches (54) while a large fruit fly (57) hovers in the air this ancient landscape with another group of butterflies, the brush-footed or nymphalid butterflies (53B).

Not all caterpillars from that period belonged to butterflies and moths. There were also argid sawflies, the larvae of which resemble those of moths and butterflies.[38] Sawflies are plant-feeding wasps, and argid males are characterized by the structure of their antennae (54). They appear to possess four antennae whereas only two are found throughout the insect world. A closer look reveals that each antenna is forked at its base, giving the impression of double the normal number.

If one were looking into the canopy layer of the amber forest, one of the twigs might move, even in the absence of a breeze. A closer look would reveal not a twig but a phasmid or stick insect with a pair of protruding eyes and six spindly legs

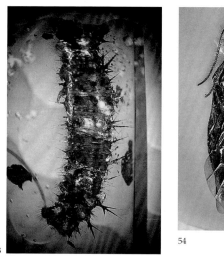

53B

54

53B. This caterpillar possesses branched spines characteristic of the brush-footed butterflies (Nymphalidae: Lepidoptera). Near the back of this larva is a hair tentatively identified as that of a bat, which could have been searching for insects among the foliage.

54. This male sawfly (*Didymia poinari:* Argidae: Hymenoptera), with its two branched antennae, at first seems an anomaly among the insects (C).

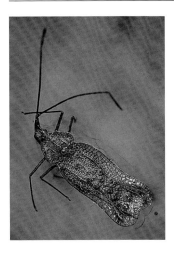

55. The cuticular patterns on lacewing bugs (Tingidae: Hemiptera) can be quite intricate (C).

emerging along the body. Two long antennae above the eyes might be slowly waving. If it were a mature female, we might see a hard-shelled, seedlike egg fall from the tip of her abdomen and land in a pool of resin. When abundant, voracious walking sticks can strip a tree of its leaves in a short period, sometimes even killing the tree by continuously consuming new leaves.

Dwelling alongside walking sticks in their former province were plant bugs flitting over the foliage surfaces, avoiding detection by darting to the opposite side of the leaves. Some of the most elegant and delicate of all insects are the lace-winged bugs (55). Just a glance at the bodies of these bugs, their reticulated surfaces resembling fine Belgian lace, explains how they received their common name. Only 3–6 mm long, these insects colonize the undersurfaces of leaves. After a while, the leaf takes on a very domesticated appearance as it becomes littered with excreta, cast skins, eggs, nymphs, and adults. This spoilage is more than compensated for by the exquisiteness of the intricate filigree patterns on the wings and head shield of these creatures.

A range of beetles would be expected in the same general habitat. From actual specimens as well as plant damage on leaf fossils, we know that leaf beetles and weevils were abundant in that ancient land. Beetles come in a variety of shapes and colors, and while the larvae are often hidden in concealed abodes such as the soil, rotting vegetable matter, and even leaf mines, some adults will feed exposed on leaves, buds, petals, and pollen.

Some beetle larvae that operate in the light of day construct protective cases to carry with them (56), ever ready to retreat into these domiciles when danger threatens.[39] A closer inspection of these cases reveals they are made up of organic debris, including even the excrement of the bearer. This is taking to the extreme the concept of recycling everything you eat. As an alternative to protective cases, other larvae have grown camouflaging hairs on their backs giving the impression, as they crawl along, that they are only a bit of debris or a portion of a lichen (see figure 92).

Huge clouds of small flies, the gall midges and moth flies, must have constantly drifted and swirled around the tree bark since their fossils are so common. The gall midges, most of which are much smaller than mosquitoes, are known to feed upon a wide range of plants, from herbs to trees, often causing swellings, or galls, on various plant parts. The larvae of some species are even able to feed on tree resin. A moth fly ensnared in a spider web is depicted in amber (see figure 70).

More robust flies darted among the tree trunks and branches, among them forms with black-spotted wings and large eyes. These large fruit flies (57) also dined on plants

56. Beetle larvae also construct protective cases, often using their own waste material, as depicted with this leaf beetle (Chrysomelidae: Coleoptera).

57. Large fruit flies (near the genus *Ceratodacus:* Tephritidae: Diptera) often have striking wing patterns (C).

58. A flat-footed or platypodid beetle of the genus *Cenocephalus* (Platypodidae: Coleoptera) rests adjacent to a cylinder of compressed frass ejected from its gallery in the trunk of the algarrobo.

throughout the forest layers, the young either forming galls, mining leaves, or consuming fruits. Those that tunneled in fresh fruit may have carried with them symbiotic bacteria that first broke down the fleshy portion, thus enabling the young larvae to use the nutrients for their development. Such associations occur with present-day relatives of this group such as the apple maggot and olive fly, well known for their destructive habits.[40]

A number of insects excavated bark and wood of the trunks and branches, remaining concealed most of their lives—exposed only when leaving their old home or entering a new one. As indicated by the frequent occurrence of fossils, flat-footed beetles must have thrived throughout the original timber stands.[41] These elongated cylindrical beetles tunnel deep in the sapwood and hardwood of older trees, especially weakened ones. They deposit their eggs in loose clusters in the passageways and work hard to keep them clean—constantly pushing the wood particles (frass) out of the small entrance holes. These masses of frass often assumed the cylindrical shape of the gallery and together with the beetles (58) landed

A scene depicting life on a trunk with two circular holes made by flat-footed beetles, one of which has just pushed a cylindrical bole of frass (photo 58) from its gallery. Waiting to grasp onto one of the beetles for transport to a new habitat is a pseudoscorpion (60). A small predatory mite (65) adjacent to the beetle hole may itself be devoured by an approaching harvestman (75).

in the resin when it flowed through the beetle's tunnel, washing out both, possibly in a cleansing action. Platypodids carry on their bodies, and introduce into their dwellings, a type of fungus known as ambrosia.[42] The word "ambrosia" denotes a food that is extremely tasty and pleasing—and to the young platypodid larvae, it is exactly that. The fungus supplies essential nutrients needed for the successful development of larvae, and, for this reason, spores are carried on the adults to be introduced into new galleries. Note the backward-pointing spines on the end of these beetles. When a predator enters the beetle's passageway, the beetle backs up and wedges its body against the sides. The intruder is met with a spine-encrusted door that prohibits further entry.[43] Sharing the platypodid's domiciles are mites that feed on the ambrosia fungi, as well as pseudoscorpions that feed on the mites. Both housemates utilize the beetles to carry them from tree to tree. Although mites are difficult to see attached to the underside of these beetles (59), the pseudoscorpions (60) are quite conspicuous, usually hanging on to the rear end of the beetle by a single pedipalp. It is quite likely that this and other insect-pseudoscorpion phoretic associations are obligatory for the pseudoscorpion.

59. A mass of phoretic mites (hypopodial deutonymphs: Astigmata: Acari) hoping to be carried as a cluster on the underside of a platypodid beetle. These mites attach to their carrier by small suckers on their undersides (C).

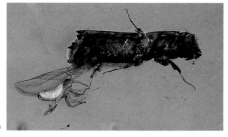

60

61

62

60. Pseudoscorpions habitually cling on insects like this platypodid beetle to reach new habitats. Some phoretic nematodes (seen below the beetle) were also hitching a ride (C).

61. Feeding on the inner bark, this bark beetle (*Cladoctonus angustostriatus:* Scolytidae: Coleoptera) emerges from its somber home to reproduce (C).

62. Long-horned beetles (Cerambycidae: Coleoptera), with their extended antennae, represent some of the longest beetles known (C).

Less conspicuous than platypodids are bark beetles, or scolytids, that feed on the bark of trees and shrubs.[44] The specimen shown here (61) belongs to a group that burrows into the living phloem layer of the host plant and forms grooves (some can form quite extensive patterns) under the bark. These are related to the Dutch Elm beetle, notorious for carrying the Dutch Elm disease fungus.

Another group of beetles, usually much more massive than the platypodids and bark beetles are the long-horned beetles (62). The common name refers to their long, slender antennae, which are usually incomplete in fossils, portions having been polished away. Members of this group have the greatest length of all known beetles, and their size may be due to their

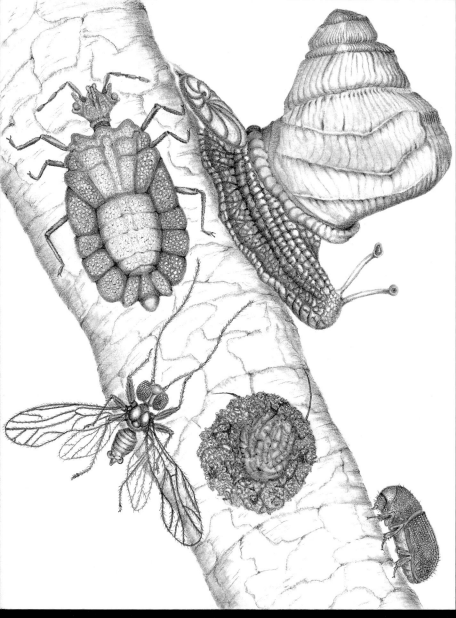

Near the base of a small sapling occur other bark and litter forms, including a snail carrying its operculum (photo 85). Resting on the bark is a wingless flat bug (82), and close behind are two bark lice, one a fully winged adult (90); the other is undetectable unless you are able to notice the two partial antennae sticking out from the circular pile of debris (91). In the lower right corner is a small scolytid bark beetle (61).

extended maturation period. In fact, the all-time record for the developmental time of any insect belongs to a long-horned beetle larva that remained in a piece of wood for forty years![45] Of course, it probably wasn't feeding all that time. The larvae of these insects, as with other wood borers, are quite rare in amber. The mandibles of larger long horns are so immense and powerful that fantastic stories arose about how the beetles were able to cut tree branches by biting them and then flying around and around until the wood was cut through. Certainly some, like the four-inch-long harlequin beetle, are impressive enough to instigate such yarns. This particular beetle is also a walking, flying menagerie. Under its thick, hard outer wing covers (elytra) and inner membranous flying wings can reside several types of lodgers. One contingent are mites that may feed on fungi and bacteria on the back of the beetle and certainly use the beetle to carry them from one location to another. It has been speculated that the mites might aid the beetles by ingesting potential microbial pathogens. Several species of pseudoscorpions have also been reported from the backs of these huge beetles. Some of them dine on the mites while others just want to reach a new habitat or are looking for a mate. Some large males remain in this location and mate with females when they board the beetle.[46] It is possible that some pseudoscorpions remain on these beetles for weeks subsisting on mites just as they do on the bark of trees or in the beetle tunnels. Perhaps the mites seek to escape these predators by crawling into small depressions on the back of the beetle called mite pits. During flight, the beetle carries its menagerie with it. To the inhabitants, it must be like living on an airliner that has regular departures and arrivals.

A very exotic, rare insect (members of this group are uncommon today) is the timber beetle (63). Some of these long, slender insects don't look like beetles at all because they have long membranous underwings that stretch the length of the elongated abdomen, far past the short, stubby forewings. In fact,

63. The front wings or elytra of most beetles cover the membranous back pair. An exception is shown here with a timber beetle of the genus *Atractocerus* (Lymexylidae: Coleoptera) (C).

one has to look very closely to locate the vestigial forewings. The head consists mostly of eyes, and the mouth parts are gill-like (flabellate). Extant larvae live deep in the wood of standing and freshly fallen trees, feeding on ambrosial fungi. The fungus is intimately associated with the beetles and spores are carried in special pouches near the ovipositor of the female so when an egg is deposited next to a crack in the bark, it is covered with them. The newly hatched larva then carries the spores on its body as it starts boring into the wood, and eventually the fungus spreads to the tunnel walls, giving them a creamy appearance. The young of the timber beetle feed on the fungus and fungal-impregnated wood, obtaining enough nutrients for their development. Although there are no timber beetles today in the Dominican Republic, one does occur in Cuba[47] and this species has been recorded as feeding on cativo, one of the amber forest trees.[48] It is likely that the one shown here (63) also fed on the wood of this plant.

Symbiosis between microorganisms and insects is extremely widespread. Many arthropods that feed on plants have some type of relationship with a microbe in order to supply nutrients or break down the vegetative material into usable food. The microbe may be carried in special pouches on the outside of the body, as with the ambrosial fungi mentioned above. Alternatively they may be carried internally in the alimentary tract, as is the case of termites that depend on gut bacteria and protozoa to digest the ingested wood particles. Many blood-suckers have symbiotes in their body cavities that play a role in the breakdown of heme proteins. The location and diversity of arthropod symbiotes is quite amazing.[49]

Invertebrates that normally live in litter on the forest floor were also preserved, but we can only speculate about how they became entrapped. They may have encountered the resin at the base of the algarroba tree, or from occasional excursions to the canopy layer. The predominant inhabitants of litter and soil are mites, springtails, and nematodes.[50] Representatives of all of these also resided on trees, inside burrows of wood-boring insects, in rotting debris, or among the epiphytic plants that must have decorated the tree limbs. The rarest of these occupants are the nematodes, since microscopic round-worms have no hard parts and quickly decay.[51] The litter forms are microbotrophs, animals that subsist on bacteria, fungi, and protozoa. The only asset these defenseless crea-tures have is a rapid development time (their life cycle can be completed within a week), which is fortunate indeed since ne-matodes are a major food source for predatory mites and a delicacy on the menu of springtails.

Springtails (64) are small, six-legged creatures that derive their name from the hinged, springlike appendage attached to the tip of their abdomen and held folded beneath by a catch. When danger threatens the catch is released, the appendage whips down against the substrate, and the creature catapults into the air like a rocket. This is the closest springtails come to flying since they are wingless. Blind individuals living deep in the soil have no need of a spring and grope their way through the soil particles. Many of the surface-dwelling springtails have yet another curious appendage protruding from the un-derside of their abdomens—a glue peg which emits a sticky substance thought to hold the animal to smooth surfaces, al-though scientists have found that these pegs also absorb wa-ter. Springtails reproduce rapidly, and sometimes they gather together in masses and migrate over the soil surface, producing

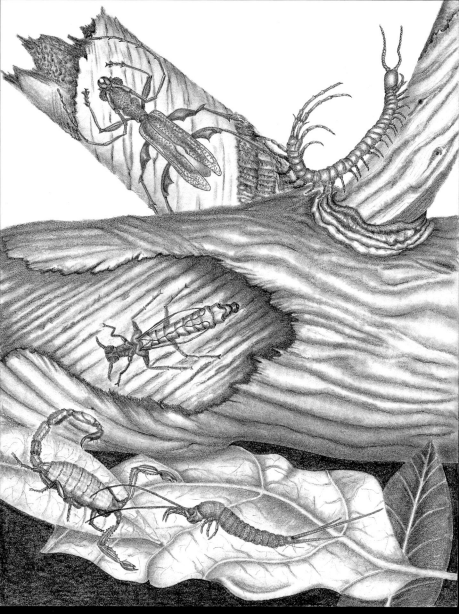

In this depiction of the litter and log zone are some of the largest arthropods captured in amber. At the top left, a longhorned beetle (photo 62) rests on a broken branch, while immediately under it is a timber beetle with its membranous hind wings (63). A centipede (87) in the upper right searches for prey, while a juvenile scorpion (76) in the lower left encounters a bristletail (83).

a seething spectacle of jumping, bubbling individuals. Mites and ants are effective predators of springtails. In fact, some ants have become specialized hunters of springtails and have evolved adaptations to grasp them before they can leap away.[52] These ravaging ants prowl over the soil surface with their mandibles held ajar by a catch that is triggered via a pair of sensory hairs protruding near the mouth opening. When these hairs touch a springtail, the mandibles snap shut like a bear trap on the hapless victim. As a defense against ants and mites, some springtails produce poisonous chemicals and assume bright warning colors. Such notifying coloration demonstrates that most predators have learning capabilities, since after attempting to eat several foul-tasting red springtails, they search for differently colored prey.

Springtails inhabit the same niches as mites. Since springtails don't eat mites, the latter seem to be the only ones that benefit. Mites are generally less than a millimeter in size and flourish almost everywhere on land and fresh water.[53] Those in the

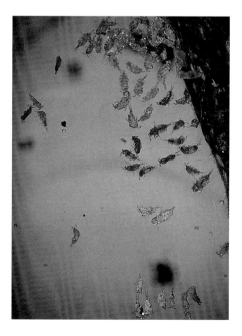

64. Springtails (Collembola) are among the most numerous arthropods in the litter zone and often travel in groups (C).

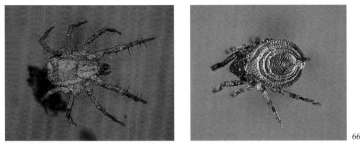

65. Free-living predatory mites like this member of the genus *Procaeculus* (Caeculidae: Acariformes: Acari) attack other small arthropods in the litter zone and use the setae on their front legs to rake in the prey (C).

66. This mite of the genus *Liodes* (Liodidae: Oribatida) demonstrates its maturity by exhibiting a dorsal set of concentric rings, each marking a molt (C).

litter layers may be plant-feeding or predaceous. One of the carnivorous forms has stout hairs on its front legs that are used to "cage" victims while feasting on them (65). Other types of mites feed on fungi and other small organisms. The one depicted here has concentric circles on its back, each indicating a past molt (66). Most mites have four pairs of legs in the adult stage but only three pairs when they hatch. Mites can be extremely abundant in soil and litter layers, sometimes reaching a million individuals in a square meter. But everything has enemies, including mites. In the ancient forest, several types of beetles such as the antlike stone beetles (67) and short-winged mold beetles (68) had a diet restricted in large part to mites, and pseudoscorpions and spiders also dine on them.

A droplet containing a microworld has been preserved in a small piece of amber. This extraordinary piece (69) contains over five hundred individuals in various stages of development. These invertebrates may belong to the phylum Tardigrada, which includes forms in aquatic as well as terrestrial habitats. Also enclosed were fungal strands upon which these arthropods were probably feeding. In addition, several predators were also present, including giant protozoa, oribatid mites, and a pseudoscorpion. Free-living nematodes probably feeding on bacteria were also there, thus showing that an entire

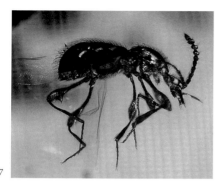

67

68

69

67. Mites are best to evade this predaceous antlike stone beetle (Scydmaenidae: Coleoptera).

68. Still another mite predator is this short-winged mold beetle (*Apharus* sp. Pselaphidae: Coleoptera).

69. These extremely small arthropods could be tardigrades (Heterotardigrada: Tardigrada) and occur together with fungi and algae, which probably served as nourishment, as well as with predators in the form of giant amoebae, mites, and a pseudoscorpion (C).

microcosm can be gleaned from one sample. Free-living nematodes have been found in a number of fossil pieces, especially those containing frass of wood-boring beetles. Many of these nematodes form resistant juvenile stages that attach to various insects for rides to new and better locations.[54] See photo (60) for some phoretic nematodes that freed themselves from a platypodid beetle. Similar nematodes carried by ants also occur in amber.

A plethora of spiders probably populated all levels of the forest. Jumping spiders frequented the litter and plant surfaces. These fearsome predators crawl quietly or simply wait for insects to appear. Their leap is so rapid as to be almost

70. This moth fly (Psychodidae: Diptera) made a wrong turn and ended up in a spider web made by a member of the family Araneidae (Araneae) (C).

71. An ant (*Azteca alpha:* Formicidae: Hymenoptera) was struggling valiantly to escape from both a spiderweb and the sticky resin (C).

undetectable. One moment the spider is resting on a dead leaf and the next it is subduing its prey several inches away. Although jumping spiders normally don't make webs, most spiders do, from silk produced in abdominal glands.[55] The silk, which starts out as a liquid, issues as fine strands through nipple-like projections called "spinnerets," usually located at the tip of the abdomen. Each type of spider makes a characteristic web although a basic pattern is usually followed. The radiating spokes of the web are made from dry silk, while the circular ones are composed of sticky strands. The spider walks on the dry strands and the victim is entrapped on the sticky strands. Some resourceful insects have learned this trick and live in spider webs without being caught, dining on the carcasses after the spider has finished. Some webs, even with their trapped victims, have been found in amber. In fact, the web containing the entrapped moth fly (70) is so well preserved that

72. Many female spiders, such as this member of the family Pholcidae (Araneae), carry their eggs to protect them (C).

73. Young spiders like these (Araneae) may have been preparing to balloon, a process that allows spiders to fly by emitting silk strands that are carried by the wind (C).

it is possible to identify the maker as a member of the spider family Araneidae. In a sense, the timid moth fly and the pugnacious ant (71) were both victims of double jeopardy, being caught twice. While the moth fly appears to have expired quietly, the ant perished struggling; her attempts to escape are clearly depicted in the photograph.

Silk is not solely employed for making webs. It is also utilized for constructing egg cases, which are often carried around by the spider. This long-legged spider still holds her egg sac with a thin strand of silk attached to her mouthparts (72). Also caught were recently hatched spiderlings (73) that may have been preparing to float into the air when the resin entrapped them. Ballooning, the closest that spiders come to flying, is initiated when the arachnids crawl to an exposed location and emit silken threads from their spinnerets.[56] When the filaments reach a certain length, depending on the wind velocity, the spiderlings are lifted into the air and can be transported long

74. A spider of the genus *Miagrammopes* (Uloboridae: Araneae) uses its front legs to tension-feel the presence of prey in its web (C).

distances before being dropped to earth again. Some spiders have been collected thousands of feet in the atmosphere, demonstrating the effectiveness of ballooning in distributing these creatures.

Although spiders are themselves very effective predators, there are some wasps and flies that rear their young on them, so they are very much on the defensive. Some have developed ways of camouflaging themselves on their webs by appearing as bird droppings or as twisted leaves.[57] Others actually cover their bodies with the remains of their victims. A few take on the appearance of ants or wasps which tend to be distasteful and are avoided by predators.

An uloborid spider (74) developed an ingenious manner of catching its prey. After constructing a web it waits off to one side, keeping its front legs in contact with several silk strands connected to the web. When a victim becomes entangled, the change in tension signals its presence and the spider rushes to the attack. Clever individuals have learned to discern between the vibrations of defenseless prey and spider wasps that imitate prey by pulling on the web to entice the spider out of hiding.

Another group of arachnid predators in the litter zone of this prehistoric world were the harvestmen. Some of these forms

75. A female harvestman (*Hummelinckiolus silhavyi*: Phalangodidae: Opiliones), of minute proportions, patrolled bark crevices for prey (C).

76. A scorpion (Buthidae: Scorpiones), still assuming its defensive position, with open claws and poised stinger (C).

were unpretentious (75) and restricted their quarry to lesser spiders and mites. These harvestmen hunted on tree trunks under overhanging bark pieces as well as in the litter on the forest floor.[58]

One of the rarest and most sought-after fossil arachnids are scorpions (76).[59] These arachnids have their pedipalps greatly enlarged with formidable pincers on the tips. The pincers are employed in defense, but their main use is to capture and hold prey. Also enlisted in subduing a victim is poison emitted from a stinger on the tip of the scorpion's tail. Most scorpions hunt at night and include in their diet almost anything they can capture, including small vertebrates.[60] Scorpions have reduced eyes and poor vision, apparently detecting their quarry by feel and possibly aided by a pair of comblike organs called pectins located on the undersurface of the body. For some unknown reason, many scorpions fluoresce under ultraviolet light and collectors take advantage of this by using blacklights to detect their presence. Nocturnal scorpions, hunting in the filtered moonlight, will scale tree trunks in search of food. Others reside in protected areas on the branches of trees, especially in the leaf bases of epiphytic plants. An interesting habit of scorpions and one that probably accounts for most of the preserved

Inhabitants of the litter zone included a forest hopper (photo 81) that avoided
its enemies by jumping. Beneath it is a wood louse. These crustaceans as well as
the millipede (88) were important detritivors in the amber forest.

77. One normally thinks of wind scorpions as running over desert sands, but some, like this one (*Happlodontus proterus:* Ammotrechidae: Solpugida), coursed up and down tree trunks in search of prey (C).

78. Tail-less whipscorpions (*Phrynus:* Amblypygi: Arachnida) represented some of the largest arachnids in the amber forest (C).

specimens involves their young. Born fully formed, they ride on their mother's back for up to a month until the first molt is completed. Many amber scorpions are juveniles that probably slipped into the resin during the nightly hunting forays of their parent. Although scorpions appear so formidable that one would suspect nothing would dare approach them, they are attacked by large spiders, centipedes, other scorpions, and even vertebrates.

Other large exotic arachnids in the amber forest were wind scorpions and amblypigids. It was a surprise to discover entombed wind scorpions (77), since these fleet creatures are usually associated with dry, desert climates.[61] However, some wind scorpions do live in tropical forests today, scampering up and down tree trunks in search of a meal; they hold the record as the fastest known invertebrate. Tail-less whipscorpions or amblypigids (78) are also efficient predators, remaining hidden under rocks or bark most of the day, coming out only at night to feed. Their chelicerae are modified into robust, spine-armored

79. Pseudoscorpions (Pseudoscorpiones) resemble miniature scorpions with their large pincers, but they lack a terminal stinger (C).

80. Woodlice such as this one (Pseudarmadillidae: Isopoda: Crustacea) are important detritivores that look like miniature armadillos (C).

grasping organs that hapless arthropods would find nearly inescapable. Their front pair of legs, in contrast, are long and slender, obviously modified for sensory functions. Lacking any type of tail appendage, these formidable creatures are not frequently encountered. The females carry their eggs in a sac attached to the undersurface of the abdomen by a few silken threads.

Found in the same general habitat as scorpions and amblypigids are false or pseudoscorpions (79). Pseudoscorpions have anterior pincers like scorpions but lack the elongated portion of the tail with the stinger. Thus they subdue their game only with their pinchers. Pseudoscorpions are generally quite inconspicuous, remaining concealed under bark as well as in the litter zone. They have the ability to produce silk and spin sericeous cocoons containing their eggs. We have already mentioned the psuedoscorpion's habit of clasping onto insects, especially platypodid beetles, or climbing on the backs of longhorned beetles for transport to new localities. They are able to pick up minuscule, delicate mites and even nematodes with their pincers, which in some are drawn out at the tips like a pair of forceps.

Also in the litter zone were terrestrial crustaceans such as wood lice or isopods (80)[62] and forest hoppers or amphipods (81).[63] These two left an aquatic habitat and became adapted

81. Beaches are not the only habitat for amphipods. Many, like this forest hopper (*Tethorchestia palaeorchestes:* Talitridae: Amphipoda: Crustacea), occur in tropical woodlands (C).

for survival on land, although they still retain gill-like appendages that must be kept moist, limiting them to humid habitats. Their seven pairs of legs resemble those of shrimp, as do the armored plates that cover their bodies. Both wood lice and forest hoppers feed on living or dead plant material and have taken up abodes among the epiphytic plants and rich humus on the branches of trees. Some isopods have the ability to curl up in a ball when danger threatens, much like the armadillo. Others are covered with sculptured armor, and many of those living in the dark have become light colored. Female wood lice carry their young in brood pouches located under their bodies. A large individual with a cluster of young happened to fall into a pool of resin and the whole family is now preserved eternally. As a result of their habits, which only occasionally bring them into contact with resin, forest hoppers are rare. Those that are captured may have been fleeing from some enemy when they landed in the resin. Although they don't look very appetizing to us, spiders find these crustaceans tasty, provoking a few wood lice to develop chemical defenses.

Most animals blend into their environment unless they want to call attention to themselves, which normally is not a wise idea. Some insects blend in so well that they have lost their wings and depend completely on camouflage to avoid their enemies. One such insect is a denizen of the litter zone—a flat bug (82) that took on the appearance of a useless piece of bark. Since flat bugs browse on fungi found on tree trunks, the disguise is almost perfect.

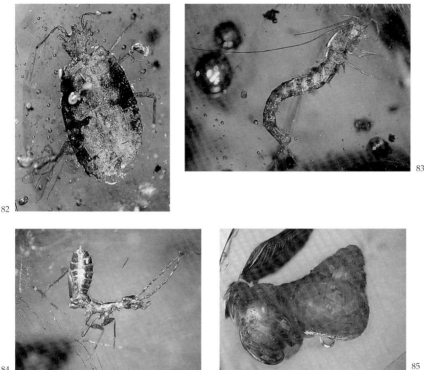

82. This wingless flat bug (*Eretmocoris* sp.: Aradidae: Hemiptera) would have been well camouflaged against bark while it fed on fungi (C).

83. Bristletails are quite primitive insects, and thus far only a single species (*Neomachilellus dominicanus:* Meinertellidae: Archaeognatha) has been found in Dominican amber (C).

84. Zorapterans conceal themselves under bark or in rotting wood and are rarely seen. This specimen (*Zorotypus palaeus:* Zorotypidae: Zoraptera) was the first fossil found of this small insect order (C).

85. A land snail (Prosobranchia: Gastropoda), with its body partly protruded and its operculum (which closed the opening of the shell) still present (C).

There are many other survival tactics. In the same general habitat as the flat bug were bristle-tails (83) that scurried and darted over exposed roots and bark.[64] These primitive, wing-less forms have limited defensive tactics (no huge mandibles, no protective case) but rely on remaining motionless and dart-ing away quickly if discovered. Yet they have managed to sur-

86. The rarity of arboreal snails like this representative of the genus *Varicella* (Oleacinidae: Gastropoda) indicates an avoidance of resin deposits (C).

87. Centipedes such as this one (Scolopendromorpha: Chilipoda) emerge at night to seek prey, which they may have to subdue with their poison fangs (C).

vive for millions of years, partly because they are generalists feeding on both plant and animal debris. Other seemingly defenseless creatures are zorapterans (84)—shy and inconspicuous, they remain under bark and sometimes occur with termites.[65] Belonging to an order of insects with fewer than thirty species, they are seldom seen today and go about their activities in obscurity, feeding on fungi and other microorganisms in the litter and bark realms.

Snails, of course, have a home to retreat into when danger threatens. In that ancient forest lived both ground-dwelling (85) and arboreal snails (86), although both types can occur on logs and branches.[66] This ground snail (85) has an operculum, a hard structure that closes the shell opening, still attached to the partially degraded soft tissues spewing from the shell. Snails feed on living plant matter; one piece of amber had evidence that a snail was feeding on a petal of the algarrobo just before both fell into the resin.

Centipedes and millipedes also crawled through the litter zone of the amber forest, although representatives of both prob-

89

88

88. Millipedes such as this representative of the order Polydesmida (Diplopoda) may play an important role in the recycling of vegetable matter in forests (C).

89. Some female millipedes (Diplopoda) tend their young after hatching (C).

ably could be found under the bark of various trees. Centipedes (87) have a single pair of legs attached to each body segment which they use alternatively as they walk, so that their bodies often wiggle from side to side. Millipedes (88) have a double pair (two on each side) on each body segment, and these appendages seem to ripple in waves as the animals move, without showing any lateral movement. The food habits of the two also vary. Centipedes are usually nocturnal and scrounge for insects, slugs, and so forth, locating them mainly by touch. Their first pair of legs are modified into poison fangs that pierce and inject venom into the victim's body. Millipedes, however, are generally detritivors that feed on decaying plant matter, moving about cautiously during the day as well as the night; a variety occur in amber.[67] Some females care for their young, and a millipede with her young has been recorded (89). Mothers may construct and then guard nests, even picking up the eggs and cleaning them with their mouthparts. Even though these forms prefer to go about their duties hidden under leaves during the day, they are still caught and devoured by birds, mice, and predatory beetles.

90. Bark lice (Psocoptera) are common in amber and must have spent much time on the bark of the algarrobo tree. This specimen has a quite striking wing venation (C).

91. A few bark lice, such as this member of the genus *Blastopsocus* (Psocidae: Psocoptera), have taken the precaution of constructing protective domiciles under which they can rest (as long as a sharp-eyed predator doesn't notice antennae emerging from what appears as a pile of debris) (C).

Another group of insects that tend to be very secretive in their habits and were well represented in the ancient forest are the bark lice, or psocids.[68] Bark lice (90) are usually solitary, but sometimes the adults and nymphs live together under webs spun out of silk emitted from head glands. They are found not only in litter but, as their name implies, on and under bark. Quite delicate and somewhat shy, many come out only at night and, in daylight, seek shelter in crevices or under fallen leaves. Some have shapes similar to beetles and enter tunnels of wood-boring insects like the timber beetle. Others have developed the ability to construct umbrella-like shelters by weaving wood particles and debris together with silk (91). When viewed from above, there is no clue that a psocid is hiding underneath, unless you can detect the tips of the antennae sticking out of the shelter! Nourishment for these diminutive insects is plentiful. Apparently they browse or scrape algae and fungi from the surface of leaves and bark, nibble on lichens, and occasionally dine on pollen and spores. These primitive creatures have few known defenses against invertebrate enemies—spiders, mites, earwigs, ground beetles, and wasps. Their best protection

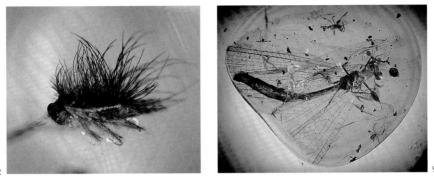

92. Some larvae, such as this beetle (Dermestidae: Coleoptera) produce their own camouflage in the form of long body hairs and look like tiny traveling forests (C).

93. This female antlion (*Porreus dominicanus:* Myrmeliontidae: Neuroptera) is one of the larger insects trapped in amber (C).

appears to be running, since they are quite agile, and hiding inside their silken webs. However, some ants are cognizant of what these webs cover and tear through them to reach the bark lice. The wing venation of some bark lice is quite intricate and reminiscent of those of lace-wing bugs. Such patterns are thought to provide camouflage by blending in with decaying leaves, a favorite habitat of these forms.

Another technique for camouflaging is to grow bristles on the body, as seen on this beetle larvae (92). Walking around with a miniature thicket on one's back may not be very comfortable, but obviously it helps to keep one alive.

A number of specialized predators prowled through this primeval abode. One, represented by only a few specimens, is the antlion (93).[69] Adults illustrate a category of insects that have no direct association with the algarrobo tree and were entangled when they accidentally flew into the resin. They probably were more abundant than their fossil record would indicate. These antlions had larvae that constructed "ant traps" in the ground which look like funnels and operate on the same principle as the pit traps entomologists construct to sample terrestrial insects moving about. The steep incline on the

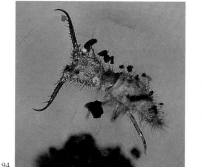

95

94

94. Curved, falcate mandibles are characteristic of antlion larvae
(Myrmeliontidae: Neuroptera) that construct pit traps in the soil to catch
wandering ants and other insects (C).

95. Velvet worms (Class Onychophora: Phylum Lobopodia) are some of the
strangest of all arthropods and have a history of over 500 million years. Slime
deposits secreted from head papillae can be seen in this specimen (C).

predator's snare makes it easy for insects to fall in but difficult
to escape, especially when the jaws of the antlion larva are open
and waiting at the bottom. They are called ant traps because
ants are so abundant on the forest floor and are much more
likely to encounter them than other insects. But the larva,
which lies partly concealed at the bottom of the pit, is not picky
and will feed on most creatures that enter. The larva (94) is
rather unique in shape and possesses long, curved mandibles
for impaling its catch. Each mandible has a channel running
through the middle which carries digestive secretions into the
victim and then transports the dissolved insect organs back
into its alimentary tract. In the rainy season, when frequent
downpours fill the pits, drenching the waiting larva, most of
the traps are constructed under overhangs or in caves. Adult
antlions are some of the largest insects preserved intact, measuring about an inch in length.

In the forest floor habitat as well as in rotting logs and even
up in the leaf bases of canopy epiphytes live some of the most
bizarre invertebrates known: the velvet worms (95), or Onychophora. They have been a source of debate since they were

discovered over a century ago—are they worms with legs, wormlike arthropods, or a missing link between arthropods and worms? Velvet worms are wormlike in appearance but have rows of legs on each side of their body. The legs are not segmented but simple outgrowths from the body wall. The animal seems to be blind, although there are two tiny beadlike structures under each antenna which are called eyes. Most intriguing are a pair of small protuberances near each antenna known as the oral papillae, from which a liquid secretion issues that partially solidifies into a slick, sticky deposit upon contact with the air. The ability to spray copious amounts of slime over long distances is just another character that makes velvet worms unequaled in the animal kingdom. The oldest velvet worms date from some 500 million years ago, from the Cambrian period when these creatures lived in the sea. Just when velvet worms became terrestrial is not known, but this rare fossil (95) shows that they were well established in the New World tropics some 15–45 million years ago.[70] Although ancient velvet worms might have been vegetarians, contemporary representatives are predators and use their slime shooting system to capture a wide range of arthropods as well as earthworms and slugs. Once found, the prey is covered with the immobilizing deposit, and then the velvet worm tears apart the captive with huge mandibles, ingesting food and slime with gusto.

Aquatic Biota

Aquatic insects also appear in amber, although infrequently. The most commonplace aquatic forms are small flies such as midges and mosquitoes. The rarest are sizable insects such as damselflies. One compelling question regarding aquatic insects concerns their breeding sites. People living in a temperate climate automatically assume that an aquatic insect must live in a standing body of water like a pond or lake, or in a running one like a stream or river. But in the tropics, another significant source of water for a wide range of aquatic forms, from diatoms to vertebrates, is water stored inside or on plants, technically called phytotelmata. While tree holes and pitcher plants are the major phytotelmata in temperate climates, the leaf bases of many tropical plants are modified to receive and retain water. Epiphytic bromeliads are one example of plants that retain water in their leaf bases. These so-called tank bromeliads can be quite large and store gallons of water, although only a few pints are needed to support an extensive selection of aquatic life.[71]

Most tank bromeliads live on the branches and trunks of other plants and appear at all levels, even in the canopy layer. We can imagine that epiphytes in the amber forest were as common as they are in present-day tropical forests. It has been estimated that from one-third to one-half of the plant species in tropical rain forests live on other plants, and many of these exist in the canopy.[72] While tank bromeliads hold standing water in their basal leaves, the leaf spaces between the upper leaves may contain wet decomposing organic matter, thus creating an additional moist microhabitat. Thus, one plant can provide refuge to a host of invertebrates and vertebrates. Numerous investigations have been made on the life forms that populate today's tropical tank bromeliads, and these can serve as a guide to indicate which of the aquatic invertebrates found could have lived in these plants in the distant past.

It is amazing how many aquatic organisms have been recovered from the water in tank bromeliads. Among the microorganisms collected are filamentous algae, diatoms, fungi, bacteria, and protozoa.[73] Diminutive multicellular animals include turbellarians, nematodes, rotifers, gastrotrichs, annelids, ostracods, copepods, leeches, and water mites. Among the insects are water bugs, beetles, flies, dragonflies, and damselflies, a single species of caddis fly and moth, and one record of a stonefly. Also included are representatives of three families of frogs. One of the most unusual inhabitants, however, are crabs that care for their young in these locations.[74] So these plants can be fairly complete microcosms.

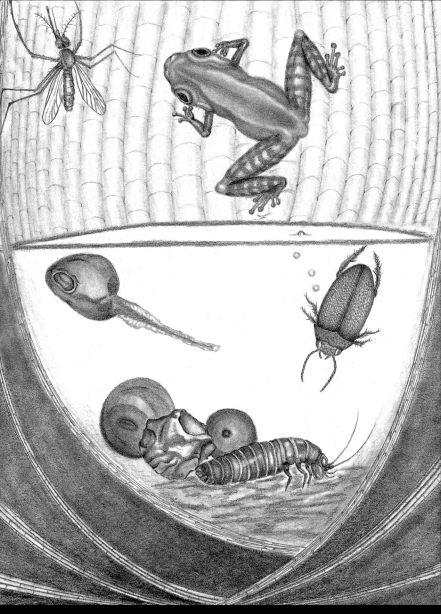

Life in water that collects between the leaves of tank bromeliads can be quite diverse, ranging from marsh beetles (photo 96) feeding on detritus and predaceous diving beetles (97) to frogs (158). Tadpoles (159) also occur in these receptacles, along with developing and trophic eggs. The mosquito resting here, which is about to be the next meal for the frog, was one of the many blood-sucking insects in the ancient forest.

96

97

96. Aquatic stages of insects such as these marsh beetle larvae (Helodidae: Coleoptera) are quite rare in amber. This pair was probably living in the water collected in a tank bromeliad (C).

97. Another aquatic insect is this diving beetle, a member of the genus *Copelatus* (Dytiscidae: Coleoptera), also associated with tank bromeliads (C).

Water in aerophytes accumulates from rain and condensation and abounds with organic matter that was blown, washed in, or left by the activities of various animals. This organic debris is broken down by primary consumers like bacteria and fungi. These are then eaten by unicellular protozoa, rotifers, and nematodes—which in turn are devoured by gastrotrichs, turbellarians, copepods, and ostracods. The insect detritivors, such as the larvae of marsh beetles (96), were followed by predatory diving beetles (97) and damselfly nymphs that prey upon them. The latter are at the top of the food chain in these microcosms; however, even they must search for cover when a crab enters the water. Ultimately, birds and mammals come to these miniature pools to fish for crabs. Naturalists have commented that the water in bromeliads is not foul, as one would expect with the accumulated waste products of all these organisms. Some believe that the bromeliad leaves secrete substances that digest and absorb the products of decomposition and leave the water fairly clear and pure.[75]

The bulk of aquatic insects found so far have relatives that breed in tank bromeliads. Certainly not all the aquatic arthropods in the ancient forest developed in bromeliads, but many probably did. The nonbiting and biting midges and mosquitoes (see figure 169) were probably very plentiful in these plant pools. Thus far, a range of mosquitoes has been recovered, including an anopheline mosquito. This is the first anopheline fossil found from the New World and shows that these potential malaria-carrying vectors were present in this ancient Dominican forest, quite possible carrying malarial-causing protozoa to birds and mammals. Members of *Anopheles* possess eggs that have characteristic floats and frills along their upper edge. Our mosquito deposited two eggs in the resin before she expired, and both are equipped with these floats and frills.[76]

Not only true aquatic species occur in bromeliads but many crane flies, moths, flower and soldier flies may have subsisted in moist, decaying vegetation between the upper leaves of the bromeliad plant. Marsh beetle larvae are the only true aquatic beetle larvae detected thus far. How they landed in the resin is unknown, but they may have been fleeing from predators, or their bromeliad abode may have been dislodged from a branch, spilling its contents into some resin. Both dragonflies and damselflies flitted over the primordial landscape. The damselflies were the most common based on the frequency of adults or partial adults (often just wing fragments) found. One preserved damselfly had a very long abdomen, similar to those her descendants use to oviposit in tank bromeliads.[77] A long abdomen allows the female to approach the water surface through the confined entrance while still hovering in the air.

Aside from the aquatic insects that develop in phytotelmata, a number of terrestrial invertebrates such as scorpions and velvet worms occur between the upper leaves of tank bromeliads today.[78] The upper leaves are not watertight and are filled with soil and organic matter. Because contemporary white worms have been reported from similar locations, a pair of mating annelids in amber might have come from this organic matter or soil.

Especially interesting are the vertebrates that are associated with water held by these air plants, including three families of frogs—the poison dart, the hylids, and the leptophylids—a few of which rear their tadpoles in these locations.[79] In some instances the mother frog returns periodically to deposit sterile or trophic eggs in the tanks containing her tadpoles. These sterile eggs appear to be a reserve food supplement to insure complete development of the tadpoles.

Additional aquatic insects provide evidence of more traditional bodies of water. Mayflies have never been reported from phytotelmata, and their presence tells us that ponds and rivers permeated that pristine landscape. Adult mayflies are exquisite with their many-veined membranous wings. They live only for a few days, and reproduction is their foremost function. After emerging, the adults engage in mating flights, some of which can be quite sizable and spectacular. Most of the mayflies in a swarm are males, some of which have an extra pair of dome-shaped dorsal eyes (98) to watch for females. As soon as a female approaches the melee, she is grabbed by a male and propelled aside for mating. The females oviposit soon afterwards, and generally lay their eggs in water. Most

98. The extended eyes on this male mayfly (Baetidae: Ephemeroptera) were probably used to detect females in mating swarms (C).

99. Larval mayflies typically molt into a subimago capable of flight before the final molt to the adult stage. Both of these stages are depicted here in this member of the genus *Careospina* (Leptophlebiidae: Ephemeroptera).

of the life cycle is spent as aquatic nymphs feeding on plant material, although some forms are predaceous. The nymphal stage, characterized by two or three long tail filaments and gills protruding from the sides of the body, can continue for one to two years. When mature, the nymph will emerge from its aquatic environment and molt into a winged form called a subimago. Although this subimago can fly, it is usually not sexually mature and in most cases must molt again before it can mate. Mayflies are unusual in being the only insects that molt after forming functional wings.[80] Fortunately, a subimago mayfly flew into a pool of resin and attempted to escape by molting to the adult stage. A record of the entire scenario is now captured (99).

An unusual group of aquatic insects represented in amber are the stoneflies.[81] Stonefly nymphs typically occur in fast-flowing streams where they live under rocks with mayflies, snails, and other aquatic invertebrates. The adults are slow-flying insects that never stray far from water. Stoneflies, the favorite food of many fish, as any fly fisherman can tell you, don't occur in Hispaniola today.

Another aquatic insect in the forgotten forest was the caddis fly (100).[82] Adults look very much like small brown moths. The

100. Caddis flies such as this male (*Cubanoptila poinari*: Glossomatidae: Trichoptera) provide indirect evidence of streams and ponds in the amber forest (C).

101. Still further evidence of slow-flowing streams or ponds is a pair of water striders (*Electrobates spinipes:* Electrobatinae: Gerridae: Hemiptera). The male (in the rear) is exhibiting mate-guarding behavior to keep other males away (C).

larvae have the singular ability to construct either silken nets to ensnare food flowing downstream or protective cases they carry around while feeding. Only one species of caddis fly is known to occur in plant reservoirs, and representatives of this family have not appeared in amber.[83] Thus, the presence of caddis flies indicates that both cascading rivers and quiet lakes formed vital parts of the habitat. The saddle case-maker caddis flies and the net-spinning caddis flies prefer fast-flowing streams that must have coursed through the trees. Each species of caddis fly tends to choose a certain type of material for constructing its case. Species of one genus may use stones, another sand grains, and a third plant material. Furthermore, some cases are straight, others are curved, and a few come in the form of a spiral.[84]

Other insects live on the surface instead of in the water, but they are still called aquatic because they are dependent on water surfaces. A well-known representative is the water strider, the medium to large sized bug that has evolved morphological adaptations on its legs that keep it from breaking through the surface film. Finding this pair of large water striders (101)[85] was a surprise, since they spend almost all of their life on open water. These long-legged predators can utilize vibrations to

102. Would you believe that the small body on the back of this female is a male broad-shouldered water strider? Discovering this pair of marine bugs (*Halovelia electrodominica*: Velidae: Hemiptera) in amber was quite a surprise. How marine insects came into resin is still a mystery.

locate their prey within seconds after it has fallen into the water. The males also dispatch "calling waves" to contact receptive females. A curious habit of the water striders is mate guarding. After pairing, the male remains next to the female for days, warding off all others. This postcopulatory mate-guarding behavior is displayed for the first time in the fossil record with this inseparable pair.

A still greater enigma is a pair of broad-shouldered water striders (102):[86] not only were they mating when they fell into the resin, but they belong to a group that today is associated with a saline habitat.[87] These are the only known maritime insects in any amber deposit. Called *Halovelia*, they are intertidal and live on protected oceanic areas such as coral reefs, mud flats, or rocky shores. When disturbed, they dart quickly into holes or depressions and can be seen scurrying over mud flats or mangrove roots at low tide. The big question is, how did littoral insects become entrapped in tree resin? Could the algarrobo trees have grown so close to the ocean shores that their resin dropped onto the unsuspecting striders? Or during a tsunami or especially high tide, did the waves sweep inland, dashing the creatures against the sticky bark of the trees? Or is it possible they were adapted to living in a more terrestrial habitat some 20–40 million years ago? The answers will proba-

103. Pygmy mole crickets (Tridactylidae: Orthoptera) are able to splash over water surfaces with the aid of "paddles" on the tips of their hind legs.

bly never be known. Another perplexing fact is *Halovelvia* bugs do not occur anywhere in the New World today. Their nearest location is Western Samoa in the Pacific Ocean, some 7,000 miles from the Dominican Republic.

How do aquatic insects become entrapped in tree resin? Aside from the pure chance of flying in the wrong direction, there is another factor that may have contributed to this phenomenon. Aquatic insects may mistake the reflective surfaces of resin pools for water deposits. Thus, responding to visual clues indicating that water is present, damselflies, diving beetles, and other insects probably plunged into the resin expecting to deposit eggs or swim away.

Some insects in amber frequented the wet shores of streams and lakes. One of the more unusual was a sand or pygmy mole cricket (103). These miniature crickets have their front legs adapted for digging burrows or pits in moist sand, often resting with only their head and forebody sticking out of the hole. They are active jumpers, leaping between land and water. Once in water, these energetic crickets swim with the aid of small paddles or plates attached to the tips of their hind legs.[88] Pygmy mole crickets shared their space with shore bugs— small, flattened insects found, as their name indicates, along the banks of ponds and streams. Shore bugs move around in spurts, alternating jumping and flying. This type of erratic movement explains how they might have ended up in their resin tomb. Their large eyes aid in detecting prey and evading enemies.

Social Insects

Of all the animals uncovered, none are more important for elu-
cidating the original biotic structure of the ancient forest than
the social insects. Broadly defined as those insects that cooper-
ate in caring for their young, have a division of labor, and over-
lap in generations, they include the ants, termites, and social
bees and wasps. Ants are the most abundant group of social insects with nearly fifty genera reported in Dominican amber, including at least six genera that are now extinct. They can tell
about the origin, evolution, distribution, behavior, and extinction of social creatures in past eons. Today, and presumably in
the ancient terrain, ants are an integral part in the ecology of
tropical forests and have formed numerous symbiotic associations with plants and other invertebrates. Their existence is in-
terlinked with the survival of numerous homopterans, butterflies, a number of nest symbionts, and a range of plants.[89]

So much could be told about the ants populating that ancient
ecosystem that it is difficult to know where to begin. Interesting
representatives have been selected from previously defined categories that include predators, secretion harvesters, seed gatherers, general scavengers, and fungus growers.

Predatory ants include both specialists and generalists. Many
patrol the litter zone, trunks, and canopy layers in search of
general arthropod prey. Others are more particular, hunting
specific arthropods or even only certain developmental stages.
Even species of army ants, whose stories inspire awe and fear,
can be quite exact in their choice of quarry, sometimes attacking
only other ants, even explicit species. Yet regardless of the target selected, their behavior of nomadic movements and temporary bivouacs is one of the most fascinating facets of social
insect life. Generalist army ants in South America and Africa
march through the forest, causing panic and wreaking havoc
among all organisms in their path. In Africa, one might hear
quite vividly the patter of their bodies falling from the leaves of
trees and shrubs, hitting the ground after searching the

branches for prey, or see the wild, fear-stricken flight of cock-roaches, grasshoppers, beetles, and lizards as they crawl or fly to escape the wake of an advancing army. The same terror strikes when columns march through the forests of South America. These ants are noted for the structural differences between workers (known as "caste polymorphism") exemplified by the size of their mandibles. The small workers possess normally contoured mandibles while the larger ones have huge, sickle-shaped structures that protrude like horns from the base of the head.[90]

Army ants of the genus *Neivamyrmex* which occur in amber[91] do not show this mandible polymorphism and tend to be more specialized in their search for food. Although their colonies may consist of hundreds of thousands of workers, they normally remain inconspicuous during their nomadic searches, with most activity occurring at night when the columns are subterranean or confined to the forest floor. They are especially fond of other ground-nesting ants, carrying the prey back to their subterranean homes under rocks or logs. Their lifestyle alternates between a migratory hunting phase of several weeks' duration and a stationary reproductive phase lasting about a month. This worker (104) was out on a raid when it

104. This army ant of the genus *Neivamyrmex* (Formicidae: Hymenoptera) appears to have been returning from a raid with a captured wasp pupa when it fell into some resin (C).

became entrapped in the resin. During the search, it attacked and raided the nest of a wasp, picking up and carrying off a pupa, which is also preserved just beneath the ant's body. Returning workers today carry captives beneath their bodies, straddling them as they walk. The position of the pair suggests that this method of transport was used in the past as well.[92]

Social wasps, which are assailed so frequently by army ants, produce a buzzing alarm sound when a swarm of these marauders approaches. This signals the wasps inside the nest to come to the surface and prepare to flee. If the attackers should reach the hive, the wasps will fly off, abandoning their young, apparently realizing it is futile to fight hoards of such voracious and persistent enemies. By leaving, they at least survive to construct another dwelling or possibly recolonize their old one after the attackers have departed. Some wasps have the foresight to purposely construct their nests in trees inhabited and protected by secretion-harvesting ants. These guardians will vigorously attack any invaders, including army ants, that attempt to climb their tree. Apparently by adopting the correct behavior, the wasps can construct their domiciles without inducing the wrath of these ants.[93] By now army ants have vanished from the Dominican Republic, indeed from the entire Greater Antilles.

Specialized predators often have some morphological modification, usually associated with the structure of their mandibles, that enables them to capture certain types of arthropods. One type of ant-specific predator from this extinct zoological garden is represented by the jaw-snapping dacetines. These tiny ants possess highly modified mandibles. Depicted here is a female dacetine that hunted minute, soft-bodied arthropods such as springtails (105). At the tip of the elongated mandibles are strongly curved apical teeth, typical of those that use the snap-trap method to capture prey. As previously described, the snap-trap method involves the release of a hair trigger and the rapid closure of the mandibles when an unsuspecting victim is encountered. Even in the queen here, the

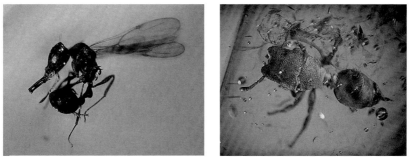

105. This queen ant (*Acanthognathus poinari*: Formicidae: Hymenoptera) with formidable bear-trap-like jaws might have been on her nuptial flight (C).

106. Workers of the genus *Cephalotes* (Formicidae: Hymenoptera) possess a shield-like head that is used to block the nest entrance (C).

mandibles have the typical elongate shape with the claws of the trap visible at the tips.[94]

Another specialization appears in some of the large cephaline ants (106). The formidable appearance of these ants evokes the idea that they must be extremely aggressive, pillaging the nests of other arthropods. But quite the opposite. The armored body of these pastoral arthropods is primarily for defense, not offense. Today a large part of the food gathered is pollen from sundry canopy trees, although they do collect dead arthropods from time to time. What has puzzled entomologists is the strange alimentary tract of these ants. There is a type of filter system that limits the ingestion of large food particles while concentrating pollen grains, which are then regurgitated as buccal pellets. Furthermore, bacteria and fungi have been discovered in the midgut portion of the alimentary tract.[95] Could these microorganisms be symbionts that assist in the breakdown and release of essential nutrients from pollen grains? Extant cephaline ants live in hollowed-out tree stems, and the modified head and prothorax are used to block the entrance of their nest. Intruders are met with a roughened shield propelled toward them with the force of six strong legs and are quickly ejected from the entrance. Since the fossils have the same morphological defensive features as their living counterparts, it

Fungus gardening ants (photo 109) conduct daily activities of collecting vegetative material for growing their fungi. Here some are ascending a tree in search of leaves, while descending workers carry leaf portions back to the nest.

109. This worker fungus gardening ant of the genus *Apterostigma* (Formicidae: Hymenoptera) was probably carrying the chewed-off leaf bit that is just above her. Back at the nest, it would have been used to nurture the fungus garden (C).

workers wind for considerable distances through the forest.[101] The workers snip small sections of leaves, which they transport back to the nest. The leaves are then masticated and prepared by other workers, who subsequently place them in the fungal gardens, often adding their own excretory secretions to the leaf compost at the same time. Larvae and adults feed on the bulbous tips of the mycelium which grow on the leaf debris.

Since these fungivores are especially fond of introduced plants,[102] they are serious pests in areas where citrus is grown. Such exotic plants apparently have no natural resistance to the ants, which seem to "know" not to take leaves of native plants that have evolved antifungal properties. Leaf cutters are an integral part of the Central and South American ecosystem in being the dominant herbivores in most areas.

Aside from the flamboyant leaf cutters, there are eight other genera of fungus-propagating ants that receive relatively little attention since their colonies are less extensive than the leaf cutters and they do not destroy living plants to obtain media on which to grow the hyphae.[103] Thus far, fossil representatives of the fungus gardeners all belong to these lesser, often considered more primitive genera.[104] The fungus gardener that we show (109) belongs to a group that today is not reported to cut

110. Harvester ants, like this member of the genus *Pogonomyrmex* (Formicidae: Hymenoptera), collect seeds. Assuming a defensive position, this worker could probably have delivered a painful sting (C).

leaves for its gardens but uses only insect feces and rotten wood. However, the edge of the leaf shown in the photo is shorn in a manner similar to leaf segments cut by true leaf-cutting ants. Whether it was cut or simply picked up by the entombed worker in the amber is difficult to say. At any rate, it is obvious that the leaf portion was being carried by the ant when it walked into the resin. Perhaps leaf cutting was a trait manifested in this extinct species in an ancient environment where true leaf cutters were absent. The nests of extant *Apterostigma* species are usually hidden in rotten wood, under rocks or, rarely, in bromeliads. No members of this genus exist in the West Indies today.

Another group represented in amber are the harvester ants (110). Their primary source of nourishment are seeds collected by the adults for the larvae, which in turn produce a liquid secretion that comprises the adult food. However, their diet can be supplemented with arthropods. Harvesting ants encompass about twenty genera worldwide. All of the individuals we have found thus far belong to the genus *Pogonomyrmex*, that still occurs in the West Indies.[105]

Apparently the retrieval of grains by harvester ant workers involves more than a random search over the countryside. They go out scouting for plants with suitable seeds, and when a good source is located, take a sample back to the nest, at the same time marking the trail with their scent (pheromone). In this way other workers can return to the original source, and if the harvest is large, columns will march out to retrieve the bounty.

Harvester ants are most abundant in dry habitats, since a dry climate favors seed preservation; in moist climates seeds may germinate in storage areas before the ants could prepare them.[106] Normally, after grains are brought into the formicary, they are stored until milled by the workers, who strip away the coat and break them apart. Some species also occur in moist and wet tropical forests, where they are most active during the dry periods. These ants can actually influence the abundance and distribution of plants. In local areas, they can lower the population density of a desirable plant below its competitive capability, thus causing its elimination. Some plants produce seeds with sweet or nutritious tips that offer the ants an additional reward for their trouble, counting on the likelihood that some of the seeds will be dropped on their way back to the nest. In such cases many of the seeds germinate—thus increasing the local distribution of a plant species. Harvester ants possess a painful sting, and in mammals the venom apparently spreads along the lymph system causing a burning sensation that continues for hours. Legends tell of prisoners being staked out over nests and left to suffer a slow, agonizing death.[107] Any mammals of prehistoric times probably gave these harvesters a wide berth.

One of the most intriguing ants in amber is *Leptomyrmex*, which today exists only in Australia , New Guinea, and some other islands in that part of the world.[108] What caused its disappearance from the Americas—its large size, or the fact that the queens of this group are wingless and, like workers, forage before a new colony is formed, thus exposing themselves to predation? Did the habit of queens establishing nests in exposed soil on the forest floor make them more vulnerable? Or was their extinction due to a climatic change, one that affected all of Central and South America and directly or indirectly led to their demise? From these fossils, we know that the workers foraged in groups and that the diet probably consisted predominantly of liquid from honeydew or extrafloral nectaries. The latter is based on the discovery of this "replete" worker (111). Some

111. This replete worker of the genus *Leptomyrmex* (Formicidae: Hymenoptera) possesses an expanded abdomen, used as a storage container for liquid nutrients. This genus of ants has disappeared from the Western Hemisphere (C).

individuals serve as reservoirs for liquid food for other ants. They have distended abdomens, nearly bursting from their burden, and sometimes become so large that they cannot walk. These living storage casks normally remain in the colony, their supply of "juice" being periodically replenished by regurgitating foragers. In the case in photo 111, it is obvious that this replete worker was able to move about to some extent at least.

A variety of diverse arthropods are associated with ants.[109] Most actually reside in the nests and obtain food by either feeding on various stages of their hosts and on other arthropods, or on refuse and corpses. In amber one finds flies, spring tails, mites, millipedes, and a range of beetles that are known today to inhabit ant nests. One of the most robust type of beetles found there is known as the "ant-nest beetle." Among the modifications these beetles have evolved for survival in formicaries are their greatly enlarged antennae. These are quite glandular and produce exudates attractive to the ants. Adult beetles have been observed feeding on both adult and immature stages of their hosts. Obviously the beetle is able to block the aggressiveness of the ants, perhaps by producing attractive secretions or by mimicking the ants. Only this could explain why the colony refrains from attacking the marauders, even as they devour young ant larvae in front of them. The ant-nest beetle depicted here (112) cannot be assigned to any definite group of ants but demonstrates the enlarged glandular antennae.[110] Its

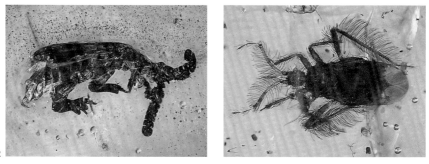

112. An ant nest beetle (*Eohomopterus poinari:* Paussinae: Carabidae: Coleoptera) possesses the typical antennal segments thought to produce attractive secretions to its ant hosts (C).

113. An ant bug (*Praecoris dominicana:* Reduviidae: Hemiptera) is covered with protective hairs to ward off attacks by victimized ants (C).

heavily armored body and appendages suggest that from time to time the ants weren't fooled and showed some aggressiveness toward it.

A variety of organisms pursue ants. Some predators have evolved unbelievable methods of securing their prey. Ant bugs (113) linger on the bark of trees, positioning themselves in the vicinity of foraging ants.[111] When hungry, an individual rears up and exposes its undersurface, revealing a gland that releases an appealing secretion. A curious ant approaches and slowly begins to imbibe the attractive fluid that issues from this unique location. At this time, the bug appears quite benign, waiting, because mixed in this appetizing secretion is a narcotic which soon affects the drinker's responses. Without warning, the bug straightens up, repositions its head, and savagely thrusts its beak into the lethargic insect's body. Although dulled, some ants respond by flaying their mandibles against the trickster. But the thick, protective hairs cloaking the bug foil most attacks and the predator sucks its victim dry. Ants bugs may be extinct in the New World today and none have been found in the West Indies.

A number of arachnids attack ants, including small *Laseola* spiders that hide in litter. The males (114) have eyes protruding

Scenes from the amber forest as viewed by a stingless bee in the last few hours of its life.

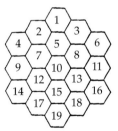

1. Scale insect (Coccoidea: Homoptera)
2. Leaf
3. Planthopper (Derbidae: Homoptera).
4. Darkling beetle (*Liodema phalacroides*: Tenebrionidae: Coleoptera).
5. Spider (Araneae: Arachnida).
6. Biting midge (Ceratopogonidae: Diptera).
7. Ant (Formicidae: Hymenoptera).
8. Snail (*Helicinidae*: Gastropoda).
9. Beetle larva (Coleoptera).
10. Flower
11. Gall midge (Cecidomyiidae: Diptera).
12. Spider (Araneae: Arachnida).
13. Winged bush cricket (Trigonidiidae: Grylloptera).
14. Web spinner (*Mesembia*: Anisembiidae: Embioptera).
15. Gnat bug (Enicocephalidae: Hemiptera).
16. Termite-nest beetle (Trichopseniinae: Staphylinidae: Coleoptera).
17. Velvet mite (Thrombidiformes: Acari).
18. Egg of stick insect (Phasmatidae: Phasmida).
19. Horsefly (Tabanidae: Diptera).

76

77

78

79

80

81

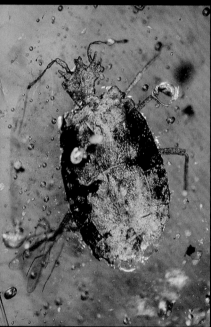

82

83

84

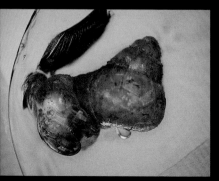

85

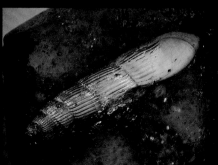

86

87

88

89

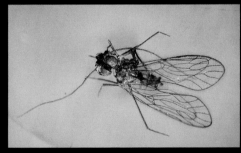

90

91

92

93

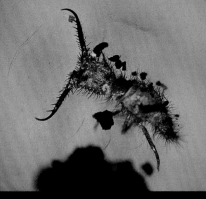

94

95

96

97

98

100

101

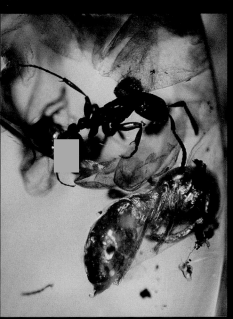

104

107

105

106

108

109

110

112

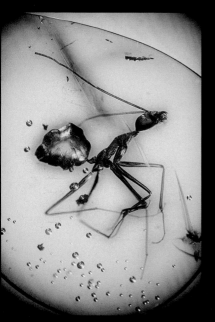

111

113

115

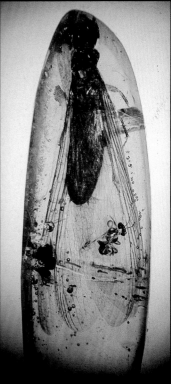

116

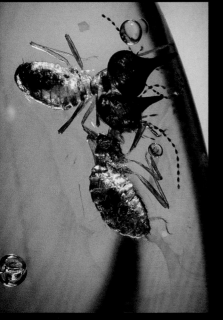

117

118

119

120

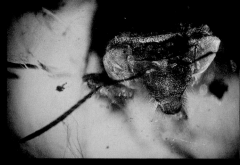

122

123

124

125

126

128

130

131

132

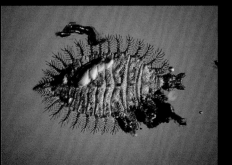

133

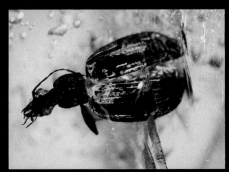

134

135

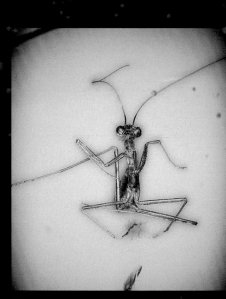

136

137

138

139

140

41

143

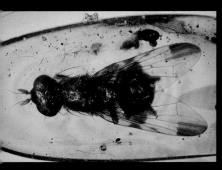

145

146

147

149

151

152

153

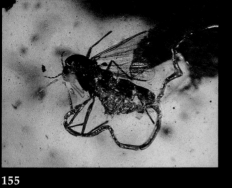

155

157A

156A

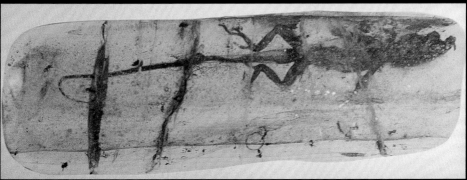

157B

158

160

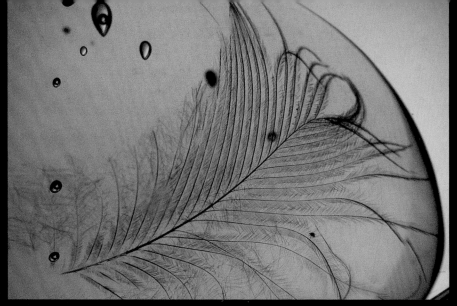

163

164

165

166

167

168

169

170

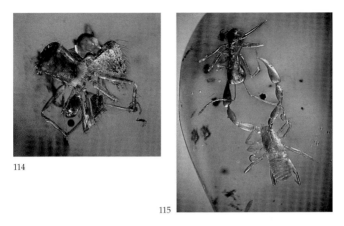

114. A male "oil can" spider (*Lasaeola:* Theridiidae: Araneae) with its characteristic cylindrical high cephalothorax dines on ants in the litter zone.

115. A pseudoscorpion and an ant (*Azteca alpha:* Formicidae: Hymenoptera) face off in an encounter that occurred on the bark of an algarrobo tree (C).

from the top of the rounded cephalothorax, perhaps to better locate the whereabouts of females. Little, if any, web is constructed by these spiders, which must then ambush their passing victims. Even while traversing the trunk of the algarrobo tree, ants are in danger of meeting predators like pseudoscorpions lying in wait under the bark. Here (115) such an encounter is frozen in time. Occasionally the ant may get the upper hand and destroy the pseudoscorpion.

There are two basic strategies for survival in the animal kingdom, exemplified by the two largest groups of social insects, the termites and the ants. In general, the latter are aggressive in nature, being equipped with the means to search and destroy, whether it be animals or plants. This assertive behavior is correlated with rapid movements and strong biting and stinging reflexes. In contrast, the former are mostly submissive in nature, and their strategies are concealment and an effective defense system. Most termites maintain themselves feeding on dead or occasionally living plant material. Each of these contrasting lifestyles is highly successful and the result of morphological and behavioral modifications that have evolved over millions of years—coupled with the ability to develop symbiotic associations with microorganisms, as epitomized in the termites.[112]

This lost world contained a multitude of termites, and the nests of many were probably quite conspicuous, especially those of the cone-nose forms constructed on the branches and trunks of various trees and bushes. Other species had domiciles inside dead logs, where they slowly consumed their abode, eventually reducing it to a powdery, rich humus. Many also excavated soil, bringing the earth to the surface for the construction of their home. These burrowing insects played an important role in aerating and supplying nutrients to the earth.

Like ants, termites have different castes that carry out specific tasks for the survival of the colony. There are winged reproductive adults, soldiers, and workers. The latter are usually very light in color, a characteristic that gave rise to the term "white ants" even though they differ from ants in both morphology and behavior. Their survival appears to be contingent on the presence of symbiotic microorganisms which reside in the gut. Termites eat wood but, as impossible as it seems, they are incapable of digesting it. This task is relegated to the protozoa and bacteria that populate the alimentary tract. If these microorganisms were removed, the insects would starve. Interestingly, these microorganisms are lost each time the host molts. After each molt, the termite must reinfect itself with these symbionts. To do this, they have developed a unique type of inoculation in which the newly molted worker obtains the symbiotic microbes from rectal discharges of its nest mates. Termites are very efficient: after the cellulose from the eaten wood is "digested" by the microorganisms, the remaining lignin is eliminated in the form of fecal pellets which are often used for colony construction.[113]

Representatives of both the so-called primitive as well as advanced groups of termites are present in amber. The most primitive of all are the giant *Mastotermes* termites, which today comprise a single species restricted to tropical areas of Australia

116. A giant termite of the genus *Mastotermes* (Mastotermitidae: Isoptera) was unable to adapt to changing conditions in the Americas, and only a single species remains in the Australasian realm today (C).

and New Guinea. Although this species seems to be thriving in Australia, much to the annoyance of humans, since its diet includes structural timbers, living trees, crops, and even synthetic materials, it is a relict. The fossil record shows that at one time the range of *Mastotermes* extended over much of the globe, including the Dominican Republic. It appears that large organisms are more prone to extinction than small ones, which certainly applies here. *Mastotermes* is the largest known extant termite and one of the most gigantic insects recovered from amber, with the sexual adults measuring up to 30 mm in length from tip of head to tip of wings (116).[114]

The living Australian *Mastotermes* construct nests in soil, logs, stumps, and even in living trees, and we can assume that the fossil forms had similar habits. Thus their colonies could have attained up to one million individuals, with queens surviving some seventeen years. The microscopic protozoa in the alimentary tracts of the Australian *Mastotermes* are quite spectacular in appearance, being endowed with parallel rows of flagella which produce shimmering waves of color.[115] It is almost certain that the ancient forms had similar symbionts, and although they actually have not been noted there is indirect evidence of their presence. Large bubbles emerge from the alimentary tracts of some entombed *Mastotermes*. These air bubbles likely represent methane gas that is known to be produced by protozoa in the gut of these insects. The microorganisms probably survived just long enough after their hosts became immersed in resin to produce this excretory product.

There are some pressing questions regarding the demised *Mastotermes:* Why and when did they become extinct? Did a new predator enter the ecosystem, one that favored the larger size of *Mastotermes* as an easy food source and against which the soldiers of these giants, even with their slashing mandibles and noxious chemicals, had little effect? Or was there a gradual change in the temperature and moisture conditions on the island and in neighboring Central and South America, which the giant termites or possibly their internal symbiotes could not tolerate?

One of the advanced groups of termites of the distant past were the nasute or cone-nosed termites, distinguished by the cone-shaped head of the soldiers. These highly evolved termites have replaced protozoa with symbiotic bacteria that play a similar role in digesting cellulose.[116] Nasute termites still occur in the Dominican Republic today—in fact, they are fairly common throughout Central and South America. They construct large, arboreal, bulbous nests made from a mixture of chewed wood and waste material. The outside layer of their domicile is quite hard and thick, while the inside is honeycombed with galleries and passageways.

A few South American cone-nosed termites are active during the day and have been observed conducting foraging marches.[117] Termites periodically travel out of the colony in search of food materials and soil particles for nest construction. Their exodus has been compared to a well-disciplined army on maneuvers. Columns of workers, four or five abreast, start marching down the tree to the ground or across the lianas to another tree, depending on a route predetermined by the scouts. The resolute columns wind their way across the forest floor, following a scent trail laid down by the scouts and further enhanced by the front workers. Along the flanks are sightless

117. A pair of "cone-nosed termites"
of the genus *Nasutitermes* (Termitidae:
Isoptera) were chemical-squirting
soldiers from an arboreal nest. These
successful termites occur throughout
the tropics today, including
Hispaniola (C).

soldiers whose heads are constructed solely for defense, lacking even mandibles (117). The soldiers follow the column as it
proceeds through the forest, guarding the advancing hoards
and clearing the path. They march along singly or stand with
their heads pointing outward as the rank and file move past
them. If searching for soil, the workers enter pre-dug tunnels in
the earth and dislodge the particles, each worker taking up one
in its mouth before returning home over the same scent trail.
The soil will be used for the repair and expansion of the domicile. On marches when the termites forage for food, they stop at
various localities and consume plant material, storing it in their
crop and then returning home. This nourishment is eventually
divided among the colony members.

To other animals, these columns are a moving feast, with ants
being among the first to partake. Battles between ants and termites probably have been going on since both groups met back
in the Cretaceous some 100 million years ago, far longer than
the conflict between wolves and caribou or lions and hyenas.
The struggle is rather one-sided, however, with the ants usually serving as the aggressors and termites as the prey.[118] Ants
have many attack tactics, but in this case individuals follow the
columns, seizing the opportune moment to rush in and grab
one of the worker termites before the soldiers can come to its
defense. However, when a massive onslaught is directed

A high level of activity centered around the resin deposits on the algarrobo tree. A worker stingless bee (photo 123) heads back to the nest after filling her corbiculae with resin for nest construction. A soldier termite (117) that served as a guard for foraging columns of workers became entangled in the sticky material and, in dying, dispelled a defensive chemical from its pointed mouthparts. An unfortunate stingless bee has been ambushed by a resin bug (124), which has already plunged its beak into the hapless victim. Meanwhile, the owl fly larva on the right (128), partially concealed by debris with only its formidable mandibles exposed, waits motionless on the bark for unsuspecting prey.

against the columns, soldiers charge to the rescue. They scurry forward and collect at the area where the multitudes are attacking, establishing a type of wedge formation with the tip pointed toward the ants. At the moment a soldier termite detects the presence of an enemy with its antennae, it discharges a jet of fluid from the small opening at the tip of its "nose." This liquid hardens as soon as it makes contact with the air and forms a sticky glue that essentially disables the attacker. This method of defense appears to be successful against insects their own size, but large insect predators, birds, and lizards are not deterred by the sticky exudate.

The soldiers are also mobilized when the surface of the domicile is damaged. They guard the damaged area while workers hasten to bring small particles of soil, moistened with their saliva, to patch the breached wall.

Often the presence of termites can be detected only by their discarded wings, which are shed when the sexual adults land and begin to mate or construct nests. Each wing has a fracture line at its base, allowing the termite quickly to discard these appendages; thus it is difficult to find a termite with all four wings still attached to its body (118). At regular periods during the year, especially during the onset of the wet season, the amber forest must have been filled with fluttering termites. Most of these migrating adults would be eaten by various predators, but a few females would escape and locate a site for constructing a future nest. Termites then, as now, represented important decomposers in the tropical forest, breaking down dead or dying trees and recycling valuable nutrients.

118. It is unusual to find a queen termite in amber with all of its wings still attached. Some have usually broken off at the fold near the wing base, which is where they normally detach after the mating flight (C).

119. A termite bug (*Termitaradus:* Termitaradidae: Hemiptera) displays a strong flattening of its body, used as a shield to protect the head and appendages (C).

As with ants, termite nests contain an assortment of different arthropods adapted to living in these confined habitats. Some, like the termite bug illustrated in photo 119, live nowhere else except in termite chambers and are strongly modified for this existence. Their extremely flattened body is expanded along the edges to conceal the head and legs. Viewed from the top, these insects appear as miniature chitons attached to marine rocks. However, despite their appearances, termite bugs can move rapidly and, if disturbed, will immediately crouch down and attach themselves tightly against the surface. In this position, they are extremely difficult to move or pry from the substrate, as are chitons. Their bodies are covered with stubby, glandular setae which probably serve some purpose relative to their relationship with termites. The food habits of termite bugs are unknown although, because of their long and slender mouthparts, they are thought to feed on fungi.[119]

The vanished forest must have teemed with various kinds of bees, but since most did not visit the algarrobo tree, they never became entrapped in resin. The small selection of the original bee fauna preserved still represents the different degrees of sociality found in bees today.[120] The most primitive bee found relative to its morphology and degree of sociality is *Chilicola*, a small, nearly hairless bee that looks like a wasp.[121] Scientists believe that bees evolved from solitary wasps, and thus the "primitive" bees resemble them in many ways. *Chilicola* also lacks any external means of carrying pollen back to the nest.

120. This wasp-appearing male bee (*Chilicola gracilis:* Colletidae: Hymenoptera) belongs to a group that carries pollen and nectar together in the alimentary tract (C).

There are no special hair masses (called scopae) on the legs or under the abdomen that can hold the grains, nor are there flattened areas on the hind legs (called corbiculae) that are used by other bees to carry pollen accumulations. Instead, it is transported internally in the crop or the anterior portion of the digestive system. Since the bee also carries nectar in the same organ, they become blended together. This combination is regurgitated into larval brood cells by the female, who then deposits an egg on the mixture and covers up the cell. The larva develops on the pollen-nectar mixture and emerges after the mother has left the area. This type of behavior, where all the work of nest construction and provision is done by a single female without any assistance from others of her species is referred to as a "solitary" lifestyle. Males, one of which is shown in photo 120, resemble the females in most of the external characters but do not transport pollen nor participate in rearing the young, and they probably die soon after mating. Present-day members of *Chilicola* construct nests inside small stems or in preexisting cavities or burrows. The cells in which the larvae develop are produced by secretions from the female. Little is known regarding the types of flowers visited by extant *Chilicola* or whether they have formed specific associations with certain plants.

The next strata of bees captured in amber are sweat bees, or halictids.[122] While many are solitary, some show the first stages of sociality by working together. In contrast to *Chilicola*, which nests in preformed cavities, the sweat bees dig their own

121. A female orchid bee (Euglossinae: Apidae: Hymenoptera), rarer than lizards and scorpions, came to collect resin for her nest. Male orchid bees are famous for their associations with orchids, and many are the sole pollinators of these beautiful flowers.

122. Head of the female orchid bee shown in the previous figure (C).

burrows in the soil. Groups of halictids often construct their tunnels in the same localized area, and several females will often work on the same hole together. Their lairs are usually lined with water-resistant secretions produced from the bees themselves, and the individual cells are provisioned with pollen. Sometimes two or three individuals will provision cells and then take turns depositing eggs in the newly formed chambers. Several females will often work together to close the brood cells after an egg has been deposited at the base. Sweat bee females carry pollen in a mass of hairs (scopae) on their legs; in some fossils, the grains can be seen still attached to these scopae. A detailed investigation might tell us about the foraging behavior of these bees in antiquity.

A very unexpected discovery was a pair of female orchid bees (121, 122).[123] Orchid bees are large, robust insects often endowed with brilliant, metallic hues of red, blue and green. Extant orchid bees colonize tropical parts of Central and South America and are essentially solitary, never becoming completely social, but occasionally aggregating in hollow logs, tree cavities, or in the ground. While female bees visit a range of flowers to obtain nectar and pollen, the males range far and wide, especially in search of the orchid flowers from which they collect aromatic substances. It is therefore the males which gave this group of bees its name. Apparently male bees convert the aromatic material from orchid flowers into female sexual attractants.[124] From the special structural modifications that are directly related to collecting and storing these fragrant compounds, it would appear that this bee-orchid relationship is one of long standing. Many orchids are virtually dependent upon orchid bees for pollination and have evolved various devices to lead the bees to the right spot so they can receive pollinia (the special sacs containing orchid pollen) or transfer pollinia from previous flowers for cross-pollination. Thus, if orchid bees became extinct, many orchids would disappear, along with other life forms whose survival relied on these plants. This evolutionary line, where a plant that is dependent on an insect for survival uses neither food nor shelter to attract it, is unique. If male bees are dependent on fragrances to attract mates, then the presence of fossil orchid bees is another indication that orchids grew in the amber forest. Female orchid bees collect resin for nest construction, which is probably how they became entrapped. No native orchid bees exist in the West Indies today[125] and certainly their extinction must have contributed to the disappearance of orchids and other plants, as well as organisms that depended on these plants for food and shelter.

Without question, the most common bee to be found in amber is a single species of stingless bee.[126] They are so named because the stingers of these highly social bees are undeveloped and nonfunctional, and they have to rely on their mandibles to defend their nests. Today, when their domiciles are attacked, masses of stingless bees will swarm over the body of their attacker, biting everywhere and even pulling out hair. To make this onslaught even more effective, some also release caustic secretions into their biting wounds, causing a severe burning and blistering of the skin. One stingless bee was discovered adjacent to a mammalian hair, quite likely pulled out of some animal attempting to steal honey from the hive. Less combative contemporary stingless bees, suffering from both the exploitation of man and competition from introduced honey-bees, are on the verge of extinction.[127]

The fossil stingless bee was certainly highly social (as are its present-day descendants). Based on our knowledge of today's bees, we can assume that the colonies of these archaic stingless bees were probably quite large, possibly comprising thousands of workers. They probably constructed their hives in hollowed-out tree stumps, in cavities under logs or stones, or even in abandoned termite and ant nests. The opening may have been a simple hole or possibly through a waxy funnel constructed at the entrance. Portions of preserved stingless bee hives show that the construction is very similar to those of present-day species which are fabricated with a mixture of wax and resin called "cerumen." Although plant waxes are collected by these bees, they are also able to produce their own wax, which emerges as thin deposits between the abdominal segments. It is obvious that resin, especially that from the algarrobo tree, was in great demand by bees in this pristine forest. Only older workers were apparently sent resin collecting. With a temporal division of labor among stingless bees, the life of a new worker would first involve caring for the brood, making brood cells from wax, or placing the comestibles brought in by the foragers into waxen storage pots. When older, these workers would forage for pollen and nectar. When new sources were found, their locations were communicated to others by odor rather than dances. As a bee returned from a rich source of food, it would stop and deposit a secretory trail with its mandibular glands on plants, stones, or clumps of earth. This scented pathway was an effective means of guiding nest mates through the under-growth to the source. Without the continuous exposure of the sun, the honeybee's directional dances are not practical; thus in tropical forests where the sun rarely reaches beneath the canopy, odor trails are much more efficient.[128]

The final task that the oldest bees performed was resin gathering.[129] Exactly how these bees are able to collect and carry it back to the hive without getting completely covered by the

123. One of the commonest bees in amber, this worker stingless bee
(*Proplebeia dominicana:* Apidae: Hymenoptera) had already placed resin
on her hind legs and was probably ready to return to her nest (C).

adhesive material is still a mystery. But it is obvious from observations today as well as from fossils that stingless bees can
remove pieces of resin, knead them into tiny spheres, place
them on their hind legs, and carry the balls back to the nest for
construction and repairs. Collecting this viscous exudate is a
hazardous activity. Misjudging the depth or stickiness of the
deposit could lead to entrapment even before any was gathered. Then again an error in judgment might occur on takeoff,
landing the bee right into the tacky sap. Or it could be jostled
and shoved in by other bees. Many of the amber-entombed
bees had resin balls on their legs (123) indicating that they were
on their way back to the nest when some mishap occurred.

An even greater peril to stingless bees were predators lurking among the dark shadows on the algarrobo trunk. One of
the most effective hunters was (and still is) the assassin or resin
bug.[130] These large, rapacious insects skulk around the deposits
specifically to catch visiting bees, and in order to make capturing more efficient, they coat their appendages, especially their
front legs, with resin.[131] Armed with speed, sticky front legs,
and a powerful beak, they ensnare the bees as they come to collect resin. One such dramatic scene between a resin bug and a
stingless bee was frozen in time (124). The bee arrived at an algarrobo's resin deposit and began collecting. Suddenly a concealed bug came out of the shadows and launched an assault,

124. An attacking resin bug
(*Apicrenus fossilis:* Apiomerinae:
Reduviidae: Hemiptera) still has one
leg over its intended victim.
Punctured in one eye, probably by the
beak of its attacker, the hapless bee
faces her persecutor (C).

grabbing the victim with both sticky legs, then plunging its beak into the bee's head, penetrating the cuticle at the edge of the left eye (the hole can still be seen). The captive bee struggled to escape and managed to back away from its attacker. The bug lunged after it, plunging both insects deep into the resin where they perished, one leg of the attacker still over its intended victim. In a rather ironic touch to this otherwise tragic drama, the bee extruded its mouthparts or, simply put, "stuck out its tongue" in what could be interpreted as a final attempt to say the last word in this everlasting struggle.

Fossil male stingless bees are found but are very rare.[132] Although probably abundant enough in that bygone world, they do not collect pollen or resin, and their chance occurrence in amber was probably due to ill timing as they flew around nests constructed in algarrobo trees, waiting for an opportunity to mate with the emerging new queens.

Stingless bees do not occur in Hispaniola today nor anywhere in the Greater Antilles. The disappearance of these and all other bees discussed above sparks the query of what caused their extinctions in the early forest. Stingless bees undoubtedly played a crucial role in the pollination of many plants. In fact, these bees were probably what is considered a keystone species (i.e., those plants or animals that exert a substantial influence on the character and structure of an ecosystem). When a keystone species is removed from a habitat, as happened in the Dominican Republic, a chain reaction follows. In this case, the consequences would have commenced with the elimination of a range of plants that relied on this bee for pollination. With the eradication of these plants, herbivorous insects that depended on them for food disappeared. In addition, parasitic insects whose existence hinged on those herbivores would also have been exterminated.

This impact on the entire ecosystem resulting from the elimination of a keystone species is called a "cascade effect." Other likely victims of this cascade effect were probably the resin bugs, since none have been collected from Hispaniola today.

All truly social wasps belong to the family Vespidae, including yellowjackets, hornets, and paper wasps. But just as with the bees, there are sundry groups that have acquired behavior construed to be the early developmental stages of sociality. The majority of wasps are solitary, meaning that a single female constructs a nest, searches for prey to provision it, lays eggs on her paralyzed victim, and then departs. Their activities are very similar to those of solitary bees except that instead of pollen, the brood cells are stocked with animal remains.[133]

Perhaps the most primitive behavior is exemplified by roach wasps or ampulicids, representatives of which occurred in this past forest. Female ampulicids capture assorted cockroaches, stinging them and inducing paralysis. She then drags her booty to an existing depression or cavity in the soil or elsewhere, deposits an egg on the captive, and departs. There is no nest construction or mass provisioning (supplying additional prey).

A number of wasps, called sphecids, populated the ancient forest, and some of their descendants show early stages of sociality. One can assume that similar behavior occurred millions of years ago. The pemphredonids are diminutive forms of wasps, a few of which are gregarious, having many individuals constructing nests in the same local area in soil or twigs. Others even construct pear-shaped, suspended brooding chambers

125. Lesser spider wasps like this one (*Trypoxylon eucharis:* Sphecidae: Hymenoptera) collect a number of spiders for a single brood cell (C).

out of bark pieces cemented firmly together. Each nidus is com-posed of three or more cells, which will be provisioned with smaller insects such as aphids, plant hoppers, thrips, or even spring tails. These abodes are usually suspended by a stalk that is made from bits of bark and attached to plants in the shad-owy areas of the forest, usually about six feet from the ground.

Cooperative nest guarding by males takes place in various small spider wasps of the genus *Trypoxylon*, several of which were captured in resin (125).[134] Females of present-day species will first construct cells in a favorable area in the forest. Cer-tain ones will breed in existing cavities such as hollow plant stems or abandoned beetle burrows. Others construct their homes of mud and suspend them from aerial rootlets exposed beneath cut banks, or build them underneath boulders, or on the undersurface of leaves. While establishing the nursery, the females obtain nourishment from nectar and honeydew, but when almost finished, they collect other food—spiders—for their offspring. Many trips have to be made since they provi-sion each larval compartment with at least three spiders.[135]

The wasps have an assortment of techniques for discovering well-concealed spiders. When hunting, they alight on the ground and move about deliberately, tapping the surface with their antennae, flushing out hiding spiders. Sometimes they will hover, scanning the underside of branches or leaves for spiders that themselves are waiting to ambush an insect. If that fails, wasps are known to visit a cobweb and will try to entice

126. This paper wasp of the tribe Polybini (Vespidae: Hymenoptera) represents one of the few social wasps trapped in amber. This species could have constructed the nest shown in the following figure (C).

the owner out into the open by pulling at the silken strands, imitating a captured insect. In a final act of desperation, these predators will even resort to tearing open the spider's retreat (usually a funnel-shaped opening at one end of the web). No matter what method is used, once the unfortunate arachnid is located, the wasp holds it with her mandibles and front legs and stings the underside of the spider. The venom acts on the nervous system and paralyzes the prey, which is then carried back to the nest. But before this task is finished, the wasp often pauses for nourishment. Sometimes this amounts just to drinking the blood seeping from the sting wound of the victim, but at other times a hole is gnawed in the spider's side to reach its hemolymph. Helpless, the doomed creature remains alive, often exhibiting convulsive leg and mouth movements while in this state. With its movements impaired, the spider meal remains fresh until the larva has hatched and is ready to devour it. Obviously, the captive's immune system is still operable, otherwise bacteria entering the wound would quickly putrefy the host, making it unsuitable as a food source.[136]

The truly social wasps of the family Vespidae that were fossilized closely resemble existing paper wasps (126). A few of these are the architects of nests composed of clusters of open combs suspended from a plant by a thin stalk. At least one partial comb has been found thus far (127)—empty, unfortunately. Each cell in the cluster was constructed to raise an offspring. It is likely that all had departed when the nest became inundated

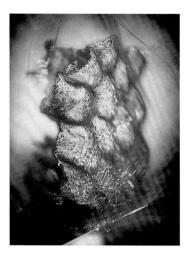

127. Portion of a social paper wasp nest, possibly constructed by the species shown in photo 126.

with resin, although the larvae could have been captured by raiding ants, one of their cardinal enemies. To further discourage ants from reaching the brood, the wasps secrete a repellent from abdominal glands and smear it on the stalk supporting the colony.[137]

Paper wasps are considered social because each breeding site is attended by at least ten female workers. One female assumes a dominant role and takes over the task of depositing eggs, eventually becoming what can be called a queen. The workers forage for insect and spider victims as well as nectar and collect wood pulp for nest construction and enlargement. Food is sliced up by the mandibles and then macerated into tissue balls before being fed to the developing larvae. Each larva dines several times before it pupates and its cell is sealed with paper pulp.

Even with the repellent covering the suspending stalk, ants still conduct raids and can quickly swarm over the colony and remove all of the larvae. The developing stages are also relished by birds and lizards. If the queen by chance is killed before she can flee, the colony dies unless another female assumes the role of egg-laying matron. In the normal course of events, though, additional females and males are formed, and the colony breaks up with young workers flying off to construct another comb.

It is interesting to note that while all of the ancient bees caught in the resin are now extinct in the West Indies,[138] most without leaving any close relatives, almost all of the higher wasps have close descendants in the Greater Antilles today. Does this mean that bees are less adaptable to changing conditions than wasps, or that predators are more resourceful than herbivores over time?

Predators

Among the diverse predators in the amber forest were members of the net-winged insects, or Neuroptera. These comprise the already mentioned antlions with larvae that construct pit traps in the soil, patiently waiting at the bottom with open jaws for an insect to slide down the sloping walls. Another member of this group with similar habits is the owl fly. While owl-fly larvae do not build pitfalls, they do bear enormous (for their size) mandibles shaped like bear traps. Instead of implanting themselves in the ground as do antlion larvae, the owl-fly larvae conceal themselves under a pile of debris, painstakingly amassed with the aid of their seemingly unwieldy jaws (128). Thus all that is visible is the deadly trap formed by the paired, motionless mandibles. When an unsuspecting insect steps between those devices, they snap shut, invariably penetrating deep into the victim's body cavity. The host juices are then transferred via a channel in each mandible to the owl fly's mouth, and the prey's dried body is then discarded.

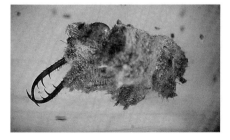

128. For its size, this owl-fly larva (Ascalaphidae: Neuroptera) supports a pair of formidable mandibles. These predators conceal their body with bits of debris (C).

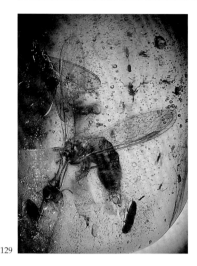

130

129

129. A mantid fly (*Mantispa* sp. Mantispidae: Neuroptera) was one of the strangest creatures in the ancient forest, grasping other insects with its mantidlike forelegs.

130. Brown lacewings (*Hemerobius* sp.: Hemerobiidae: Neuroptera) are common predators and exhibit detailed, netlike wing venation (C).

One of the strangest-appearing stalkers were the mantispids or mantid flies (129). The head of the adult mantid fly is perched at the end of the elongated thorax, giving it a sinister appearance. The adults, armed with stout raptorial front legs, can make short work of their prey. These archaic creatures flew though the algarrobo forest, catching and devouring various crickets, flies, and other insects. Young mantid flies feed on spider egg cases, first tearing a hole in the cocoons and then devouring the contents.[139]

A prevalent net-winged insect today is the brown lacewing. This specimen demonstrates the finely netted wings characteristic of this order of insects (130). Lacewings have a slow, irregular flight, and at rest they hold their wings rooflike over their back. Both adults and larvae are voracious predators of inconspicuous insects and mites, but they also partake of nectar and honeydew. The remains of this group, like all members of this order, are rare in amber.

131. Predatory thrips (Thysanoptera) are some of the smallest predators of the amber forest. Their enlarged front legs are used in grasping prey (C).

132. Dance flies (Empididae: Diptera) often have large spines on their legs for grasping small insect prey on the wing (C).

Pygmy hunters, the thrips, also abounded among the flowers. Adult thrips are fascinating with their pencil-shaped head, strange eyes, and dainty feathered wings (131). While some members of this group fed on pollen and fungal spores, others, like the one shown here (131), dined on insect eggs and unobtrusive invertebrates in the fossil environs. The body juices of the catch were sucked out with the thrips' beaklike sucking mouthparts.

Carnivorous flies were also not infrequent darting among the stands of trees. Two groups of quite effective predators were robber flies and dance flies. Today, adults of both groups are quite adept at catching insects on the wing and use their modified, spine-studded legs to imprison their quarry before they plunge their beak into the victim's body. They then release salivary secretions to liquefy the tissues and suck up the contents. A marvelous dance fly has its hind legs equipped for grasping prey prior to feeding (132).

A spider predator in the amber forest was this unusual beetle larva (133). Quite compressed, it has elegantly ornamented appendages protruding from the edges of its body. The larvae remain motionless under loose bark, waiting for spiders to

133

134

133. A beetle larva (Brachypsectridae: Coleoptera) specialized for catching spider prey (C).

134. General predators that searched the branches and leaves for prey included carabid beetles like this one (*Stenognathus* sp.: Carabidae: Coleoptera) (C).

135. Fireflies (Lampyridae: Coleoptera), shown here mating, are predaceous in both the adult and larval stages (C).

135

walk over their flattened bodies; then they suddenly swing their tail upward and impale the prey with the tail spine. Salivary juices are then discharged on the body of the victim. These beetles are not known to occur in the West Indies or in South and Central America today.

Carabid beetles are common general insect predators and the ancient forest had its share. Some, like the large one featured in photo 134, searched the trees for their meals, moving about in the day, unlike those that patrolled the ground. Another beetle predator, one with larvae that devoured insects and snails, was the firefly or lampyrid. Here is a rare example of a mating pair of fireflies (135), possibly attracted to each other by their flashes of brilliance that sparkled the evening calm of the ancient forest.

A familiar face in the bygone silva was that of a praying mantid (136). Large adults of these raptorial insects are extremely

Like most parasitoid wasps, the dryinid females will imbibe the blood of their victims. Sometimes this amounts to sipping only the fluid that oozes out of the ovipositional wound; however, in addition many sac wasps will catch and devour planthopper and leafhopper nymphs without depositing eggs in them.[144] These and other parasitoids also obtain energy from floral nectar and honeydew from various insects.

Ensign wasps are another category of fossil parasitoids. These short, stout insects receive their name from the flag like abdomen attached to the rest of their body by a short stalk. All members of this group develop in the egg cases (oothecae) of cockroaches, many of which occur in amber. The female wasps oviposit a single egg into one of the many eggs included in the cockroach oothecae. After hatching, the wasp larva devours the egg nearest it, then continues to move through the entire case, consuming those remaining before pupating in the now depleted cockroach ootheca.[145]

Two types of parasitoids studied today because of their importance in the biological control of pest insects are the braconid and ichneumonid wasps. Both thrived in the ancient biome although the braconids are the most frequent of the two and remain more common in the tropics today. Their ovipositors have been perfected, becoming long and capable of locating elusive hosts hidden in debris or in tunnels permeating wood. Instances occur where the egg is passed not through but along a small groove in the exterior portion of the ovipositor.

Thus far the importance of symbiotic relationships, and indeed how the very survival of an insect group is often based on its ability to form associations with microorganisms, has been stressed. In addition, there is an unparalleled type of symbiosis between wasps and viruses. Normally when a foreign object like an egg is inserted into the body of another insect, it illicits a defense response. Cells from the infected insect swarm over the invader and wall it off, eventually killing it. To avoid such reactions, wasp parasitoids carry viruses in their reproductive system, which they release into the host at the time of

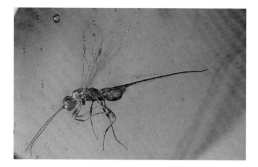

141. A male braconid wasp (*Aivalykus dominicanus:* Braconidae:
Hymenoptera), with its abdomen drawn out to a fine tip. This modification
makes it possible to mate with females even before they emerge from their
developmental site in wood (C).

oviposition. These viruses overwhelm and inhibit the host's
normally lethal defense system, at least long enough for the egg
to deceive the immune system.[146] When such a unique symbiotic system became established is unknown. These viruses are
so unique that, they have been placed in a group of their own,
the Polydnaviridae.[147] It is probable that entombed braconid
wasps also possess such viruses, one of which is shown here
(141).[148] At first glance, the wasp appears to have a long ovipositor; however, it is a male and the "ovipositor" is really the tele-
scoped and elongated terminal portion of the abdomen. The
early stages of these braconids feed on insect larvae concealed
deep within wood galleries. When the hidden female emerges
from the pupa, the male reaches in with his extended abdomen
to mate with her even before she leaves her abode. Whether the
male ever sees the female he is mating with is speculative.

Ichneumonid wasps have similar habits as braconids. The
long, slim ovipositor of many female ichneumonids, flanked
by paired oviposital sheaths, narrows to a fine point (142).
When a suitable host is detected, the ovipositor is thrust into
the underside of the insect's body. Apparently the ovipositor
tip of many ichneumonids is capable of recognizing various
host tissues, so that the wasp can deposit an egg precisely into
the nerve ganglia of its victim. Such precision may be related to

142

143

142. Ichneumonid wasps (Ichneumonidae: Hymenoptera) like this specimen were probably as common in the amber forest as they are today in the tropics.

143. A male velvet ant (*Dasymutilla* sp.: Mutillidae: Hymenoptera), probably searching for a female (C).

survival, since once the egg is ensconced within the tissue, a lethal defense response by the host is avoided. After remaining in the ganglia for a period, the early larval stages of the parasitoid assume characteristics of the host and then undetected can enter the general body cavity and feed.

Some fairly large wasp parasitoids, such as velvet ants, have been found preserved. Their common name generally applies to the females, which are wingless and superficially resemble large, hirsute ants. Equipped with a well-developed stinger capable of rendering a powerful venom, the females enter the nests of other wasps, scrutinize the cells, and deposit their eggs on larvae ready to pupate. When the velvet ant hatches, it consumes the other brood and then forms a pupal cocoon in the host cell. The winged males (143)[149] are usually larger than the females and often carry the latter around with them during the mating period, sometimes even mating on the wing, which is called "phoretic copulation."[150]

Other interesting hymenopterans residing in the original forest were spider wasps, or pompilids. These wasps are typically

144. This female spider wasp (Pompilidae: Hymenoptera) was probably
searching for spiders when it
accidentally darted into some resin.

dark, solitary creatures (144) characterized by rapid, darting
movements, accompanied by flashes of vibrating wings as they
flit about searching for spiders. Their agitated behavior, probably useful in dodging the defensive actions of spiders, has led
some to label them as *Hymenoptera neurotica*.[151] But when stalk-
ing a group well known for their own predatory behavior, such
energetic tactics could prove useful by confusing intended victims. The swift pace is usually efficacious, and by rushing in,
quickly grabbing a spider on one of its front legs and plunging
its stinger into the victim's body, a pompilid can readily para-
lyze prey much larger than itself. Sometimes the spider remains
partially conscious but is later killed by the developing larva.
Normally, paralyzed hosts never regain consciousness and are
dragged into a cavity or pre-dug burrow with an egg deposited
on their body. They serve as the sole source of food for the
wasp young.

Many more wasp parasitoids have been found entombed
than can be discussed here. Their shear numbers emphasize
the diversity of this successful group in the past as well as the
high degree of parasitism that must have occurred in the amber forest.

The second largest collection of insect parasitoids found preserved are flies, which, with few exceptions, never developed
the sophisticated ovipositor that could penetrate the body wall

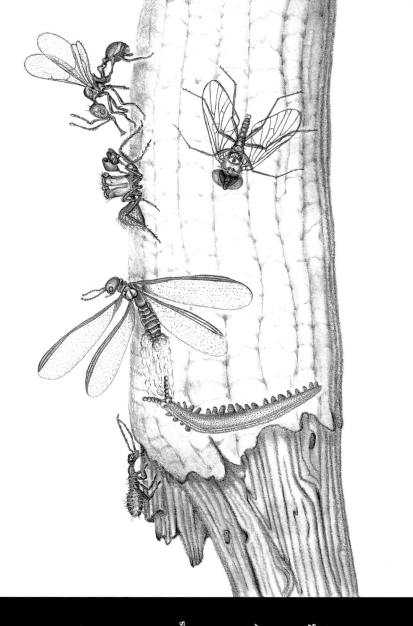

Life around a rotting log includes a velvet worm (photo 95) shooting out streams of slime from glands beneath its antennae to disable a winged termite (118) before devouring it. An "oil can" spider (114) approaches a potential meal, a small queen ant (105), which possesses elongated mandibles for capturing small arthropods like spring tails. On the log, a bee fly in the lower right (148) rests before visiting another flower for nectar and a retiring zorapteran in the upper left (84) pauses in its search for food.

145. The young of snail flies such as this one (Sciomyzidae: Diptera) are predatory on aquatic snails (C).

146. This spider fly (Acroceridae: Diptera) has huge eyes that cover most of its head, perhaps an adaptation for spotting potential hosts (C).

of their victims as the wasps did. As a consequence, these parasitoids were limited to placing their eggs on the body of their designated victim or in the close environs of the intended meal, leaving the food hunting to the newly hatched larvae. These parasitoids also have a much broader host range than wasps, attacking animals in phyla other than the arthropods.

One that attacks mollusks is a marsh or snail fly (145). The larvae search out sundry terrestrial or freshwater mollusks, usually snails but rarely slugs and even freshwater clams. In parasitoid species, the adults deposit their eggs on the shell of the snail or nearby. After hatching, the fly larva bores into the snail and begins feeding on the tissues. The host continues to survive for a while, long enough for the parasitoid to complete its development. The mature fly larva then pupates inside the shell. Larvae of some forms will devour more than one victim, assuming the role of a predator rather than a parasitoid.[152]

Another interesting group of fly parasitoids were spider flies or acrocerids (146). The one shown here has huge eyes that appear to embody all of the head, leaving little room for the small antennae and reduced mouthparts. Abundant and successful,

arachnids are the victims of a wide range of parasites, predators, and parasitoids, but spider flies appear to be anything but
aggressive. In contrast to spider wasps that dart here and there
in their rapacious search and are impossible to catch by hand,
these flies seem quite lethargic and when reposing on plants
can be easily approached and caught—hardly what one would
expect of a creature that has chosen one of the deadliest invertebrates to attack! Fortuitously, the females don't have to face
their spider hosts. Letting their larvae handle that task instead,
they deposit numerous small eggs on the ground or on plant
surfaces in the vicinity of spiders. From each egg issues a
minute larva called a planidium. They are not the typical fly
maggots that one would expect, but active, quick moving
stages that immediately start searching for spiders. Once it
makes contact with a potential host, the larva carefully crawls
up the spider's leg, moving to the back of its abdomen, conve-
niently out of leg reach. When secure, the acrocerid larva
gnaws a small hole in the spider's integument and crawls into
the safety of the body cavity, eventually taking up residence
near the host's respiratory system. The parasitoid develops
gradually inside the hapless, still-living spider, inexorably consuming the entire body contents and ultimately leaving to pupate in the web or burrow.[153]

Another, smaller fly parasitoid possessing large eyes is the
big-headed or pipunculid fly. One glance shows how they received their name (147). These are specialist parasitoids on
leafhoppers and the closely related planthoppers.[154] Female
big-headed flies are among the few that have an ovipositor capable of penetrating the integument of an insect. When a potential host is spotted, the female skims down, snatches it with
her legs, and transports the leafhopper into the air. The ovipositor is brought under the startled homopteran and forcefully
shoved through its body wall. After the egg is deposited inside,
the leafhopper is dropped and resumes its routine activities.
The fly larva is sustained by consuming the living repast, eventually emerging and pupating on a plant or dropping to the

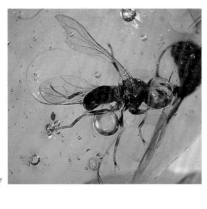

147. Big-headed flies (Pipunculidae: Diptera) have enlarged heads, as their name implies; the large eyes may assist in capturing leafhoppers (C).

148. Bee flies (*Poecilognathus* sp.: Bombyliidae: Diptera) are able to hover in midair as they sip nectar from flowers.

forest floor and pupating in the soil. Mature larvae are egg shaped and usually emerge through a perforation they produce between the abdominal segments of the leafhopper.[155]

Bee flies (148) resemble minute, fragile hummingbirds that hover with vibrating wings next to flowers while imbibing nectar with their elongated mouthparts. As they flit from one flower to another or hang poised in midair, they give scanty evidence that their larvae are parasitoids or predators. However, a female may suddenly dart down and deposit a packet of eggs near the entrance of a ground-nesting wasp or near another type of insect. Some bee flies can even shoot out eggs while hovering above the site. Once again, the larvae that hatch are slight but vigorous, actively searching for any potential meals in the vicinity. Upon contacting a wasp larva, the fly young attach to its ventral surface, puncture a small orifice in

149. Nobody knows exactly where the twisted-wing insects or strepsipterans like this *Caenocholax* sp. (Myrmecolacidae: Strepsiptera) belong in the classification of insects. The acrobatic, winged males locate and copulate with their mates inside the bodies of insect hosts. This particular male specimen probably emerged from an ant and will be attracted to its sedentary mate in an orthopteran, possibly one of the bush crickets (C).

the body wall, and dine on the fluids. After completing their development and destroying the host, the bee flies pupate in a concealed location, often within the cell of the insect they have just consumed. The mature pupa of some can be quite agile and wiggle their way upward through the soil, stopping just beneath the surface so that the adult can readily exit and take wing.[156]

Leafhoppers and planthoppers have any number of parasites to deal with. Some of the strangest are the twisted-wing parasites, or strepsipterans. Not only are these insects arresting in appearance, but their developmental habits are bizarre.[157] Scientists have been in a quandary for decades trying to resolve where these parasitoids belong in the higher scheme of classification. Only the males are winged and move freely about (149). The females almost never leave their hosts and normally are quite degenerate, lacking functional legs, wings, and eyes; just the anterior part of her body is exposed as it protrudes between the segments of the parasitized victim. This obviously presents a problem for the male. How does he mate with a female that only has her head protruding from the body of a living planthopper? Fortunately, some morphological modifications expedite this matter. The female genital opening has migrated forward and is now located behind her neck. She also emits a potent odor when she is sexually mature. This scent

attracts the male, which somehow manages to land on the mo-
bile planthopper and insert sperm into the genital pouch of the
female. Unquestionably this requires precise maneuvering and
is not without its dangers: parasitized planthoppers have been
collected with severed male genital apparati still lodged in the
female strepsipterans. After fertilization, the eggs develop
within the imprisoned female, who has little more to do in life
except function as an incubator for her young. The eggs hatch
and the young larvae migrate down to their mother's brood
chamber, just behind her neck. Collecting there, they wait for
the right signal, then pour out through the brood pouch and
disperse over the vegetation in search of a new planthopper,
which they enter by penetrating through the body wall. Fortunately, a parasitized planthopper landed in algarrobo resin just
at the precise moment that the young parasites were emerging.
Upon encountering the resin, many of these active stages
turned to climb over the body of their host. The fine appendages of these microscopic larvae, including the paired
caudal setae that assist in host penetration, can be observed
with a compound microscope (150). In addition to parasitized
planthoppers, a strepsipteran-infected sweat bee has also been
found in amber. The adult male strepsipteran figured here (149)
belongs to a group that is unique in its development: the males

150. A microscopic first stage
strepsipteran larva (Strepsiptera)
that emerged from its mother who
still resides within a parasitized
planthopper (Delphacidae:
Homoptera).

develop in ants and the females in crickets and mantids (members of the order Orthoptera). Thus a male emerging from an ant host must find and mate with a female—perhaps in one of the tree crickets discussed earlier.

Besides watching out for dryinid wasps, big-headed flies, and twisted-wing parasites, the leafhoppers have to deal with still another parasite, one that is not well known even to entomologists. This is the epipyropid moth, which has acquired a parasitic mode of life, a habit exceedingly rare for moths.[158] The adult lays her eggs on foliage upon which the leafhoppers are feeding. The hatchling is quite altered in appearance from normal caterpillars, specifically modified to ferret out hosts. After locating one, the moth larva molts and becomes quite oval in shape with six reduced legs. These parasites adhere to the underside of leafhoppers, feeding from the outside as ectoparasites. They attach themselves so closely that an indentation is produced in the integument of their quarry. Obtaining nourishment by scraping a small wound in the body wall, they slowly drain the leafhopper's blood. After reaching maturity, the larvae spin a cocoon near their spoils, usually on the plant upon which the insect host was feeding. A leafhopper parasitized by one of these moth larvae jumped into some resin and they both now lie entombed (151). The larva has released its

151. Very few entomologists have ever seen a parasitic moth larva (Epipyropidae: Lepidoptera), and this one on a leafhopper (Cicadellidae: Homoptera) was a surprise. Members of this family occur in South America today (C).

152. Parasitic beetles are uncommon, and this one (Rhipiphoridae: Coleoptera) is characterized by flabellate antennae. These beetles make a buzzing sound as they fly (C).

hold on the leafhopper, but the site where it was attached is clearly evident.

Beetle parasitoids are also uncommon today, but representatives of extant relatives have been found, such as the adult rhipiphorid beetle (152). The antennal segments of this specimen have been modified into the flat lobes (flabellate) characteristic of this group. One preferred food source of these beetles is social wasps. Again, the female does not have to encounter her quarry. Instead she deposits her eggs on or around flowers, and after hatching, the active larvae wait for wasps that alight to imbibe nectar. The first instar (stage that hatches from the egg) is quite different from the succeeding stages: it is minute, lively, and modified to search for potential meals in the environment. When the adult wasps are located, the beetle larvae crawl on their unwary bodies and are transported back to the nest. Once inside, they disembark, then seek out and penetrate the host. Transforming into more typical beetle grubs, they consume their spoils and when finished dining on the wasp larvae, the grubs pupate in the soil.[159]

A similar life cycle occurs with blister beetles. These parasitoids develop on bees, especially the ground-nesting forms. The females deposit their eggs near the nest of a bee or in an area with an abundance of flowers. The newly emerged, vigorous larvae, called "triungulins," spread through the vegetation and congregate on blossoms, then lie in ambush. When a bee arrives and begins collecting pollen or nectar, the larvae surreptitiously climb onto its back, hitching a ride to the domicile. If successful, the parasitoids are able to establish themselves in the cells, first killing and devouring the developing bees and then consuming the provisions that were laid aside for the brood. A recently hatched blister beetle larva on the back of a stingless bee was preserved (153). Since records of these

153. This blister-beetle larva (Meloidae: Coleoptera), called a triungulin, is modified to cling to the bodies of potential bees or wasps for transport back to the nest, where they feed on the young (C).

larvae developing in the cells of stingless bees today appear to be unreported,[160] either such an association was present in the past or this particular beetle larva did not realize that it had mistakenly selected the "wrong" victim. Thus, its continued development, had it not been trapped in resin, may not have been completed.

A myriad of other parasitic associations occurred in the amber forest, but only a few can be detailed here. Mites frequently obtain sustenance at the expense of insects, though often only the first larval stage of the mite is parasitic, with the remaining stages assuming the role of predators attacking small invertebrates. The six-legged larval mites usually affix to the outer surface of their victim. Inserting their mouthparts through the

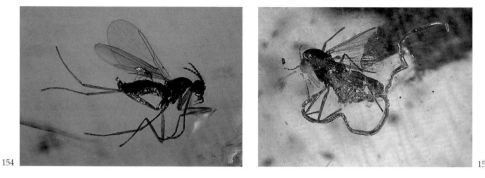

154

15[

154. Some mites of the group Thrombidiformes suck the blood of insects as the latter go about their business. This one attached itself to a fungus gnat (Sciaridae: Diptera).

155. Nematodes, especially members of the family Mermithidae (Mermithida: Nematoda), attack a variety of insects. Here two mermithids were in the process of emerging from their midge host (Chironomidae: Diptera) (C).

body wall, they slowly suck up the internal fluids, only rarely killing the host. When the mite has consumed enough nourishment to molt to the following eight-legged stage, it detaches and becomes a predator. The mouthparts of a parasitic mite can be seen still inserted into the body of this fungus gnat (154).

Another group of invertebrates that thrived in the algarrobo forest were the nematodes. While today and certainly in antiquity these roundworms occurred in a multiplicity of habitats, obtaining nourishment from microorganisms, plants, invertebrates, and vertebrates,[161] there are only a few records in amber. These include nematodes associated with moss, rotting wood, debris, and plant roots as well as two families of insect parasites. Mermithid nematodes can be called parasitoids since they complete their juvenile development in a single host, killing it upon emergence. This adult midge (155) was invaded in its larval stage by two preparasitic mermithid nematodes that obtained most of their nutriments from the midge larva. These parasitoids were then carried into the pupal and adult stages of the midge and were probably nearly full-grown when the host landed in a pool of resin. Normally, these nematodes would have emerged when the midge was near a source of water, possibly a tank bromeliad. There they would have molted into adults, then mated and oviposited. The shock of becoming inundated by resin stimulated their untimely emergence, and they were able to crawl more than halfway out of the insect before perishing. Extant midges are still parasitized by mermithid nematodes.[162] In fact, this relationship has been reported in Lebanese amber as old as 130 million years, showing that some present-day parasitic associations are extremely ancient.[163]

Another group of fossil nematode parasites discovered were the allantonematids.[164] Here, infection is initiated by a fertilized female dwelling in the environment of the target insect. This free-living adult stage bores through the larval insect's cuticle to reach the nutrient-rich body cavity. Once inside the victim's body, she swells up and produces eggs. Her progeny complete their growth, mate, and produce still another generation, all within the body cavity of the still-surviving insect. Here a drosophilid fly was parasitized by one of these allantonematids (156A). The nematodes had completed two generations within the body cavity of the fly and were preparing to emerge from the intestine and reproductive tract. Exodus was triggered as soon as the drosophilid fell into the resin. The doomed insect valiantly continued to struggle forward, leaving behind a trail of worms. Drosophilid flies are still parasitized by allantonematid nematodes in North America today.[165]

A very distinctive parasitoid that has never been definitely identified in the fossil record is a hairworm, a unique type of parasitic worm belonging to a separate phylum, the Nematomorpha. Adult hairworms are aquatic and occur in a variety of freshwater habitats. The eggs are deposited in the water and hatch into microscopic larvae, which are ingested by a range of aquatic insects. The parasites rarely develop in these aquatic insects but wait in the hope that their carrier host will be ingested by a larger, usually terrestrial arthropod. In the present case, this arthropod was a cockroach. The parasites initiated development in the body cavity of the roach, eventually reaching full maturity. In the rare specimen shown here (156B), two

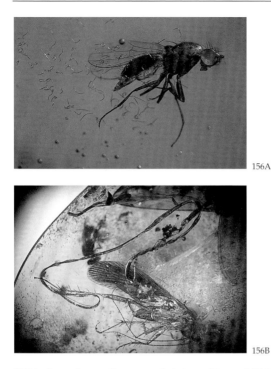

156A

156B

156A. Spread out adjacent to their host (Drosophilidae: Diptera), these nematodes of the genus *Parasitylenchus* (Allantonematidae: Tylenchida) normally leave their insect host and mate in the substrate. Then the fertilized females would have entered the body cavities of fruit fly larvae to initiate additional cycles of parasitism (C).

156B. Two hairworms (phylum Nematomorpha) are in the process of leaving their cockroach host. The posterior portion of one of the parasites is still inside the body of the insect while the other one has exited completely.

hairworms made their exit after their host became immersed in a mass of resin. One of the parasitoids exited completely, while the second was in the act of emerging through the anus of the roach when it expired.

There exists a generous assortment of microorganisms capable of causing insect diseases, including representatives of bacteria, fungi, viruses, protozoa, and rickettsias. These infective agents thrive in the internal tissues of their arthropod hosts, and most show no obvious external symptoms,

157A. Twisted strands of a
fungus grow out of the body of
this diseased bark louse
(Psocoptera) (C).

especially in the early stage of infection. The exceptions are
fungal pathogens that normally produce sporulating structures on the exterior of insects, and at this stage, the disease is
readily detectable.[166] Here is a case of one on a bark louse
(157A) where the fungal strands that emerge from the body
twisted to form a type of rope or synnematium. Other disease-
causing fungi in amber have been identified attacking ants and
termites.[167] Certainly additional insect pathogens existed then,
but there is no way of detecting their presence unless microscopic sections are made through the entombed insects. Amber
has been successfully sectioned with a microtome using the
method of double embedding to keep the fossilized resin from
shattering.[168] Further studies along this line could reveal some
interesting symbiotic associations, which up to now have only
been detected by gas bubbles emerging from the bodies of entrapped insects. Such samples of entrapped air can represent
the respiratory products of mutualistic microbes which were
living inside the gut of various arthropods, especially termites
and beetles.

Vertebrates

A moderate community of vertebrates probably lived in the
amber forest, but finding evidence of their presence has been
difficult. We can determine the existence of vertebrates using
either direct or indirect means. Direct evidence is more
compelling, although even with such material, it is often not

feasible to completely identify individual animals. Indirect data include other fossils that were dependent for their survival on vertebrates and therefore offer reliable documentation, although such signs may be indicative of more than one vertebrate group.

Immediate confirmation can be established when the entire organism is conserved or when only fragments remain. Frogs are the only amphibians that have been recovered so far, and all of the adults appear to be leptodactylids of the genus *Eleutherodactylus*.[169] These amphibians have what is called "direct development," signifying that the tadpole stage dwells inside the egg and then hatches as a minute froglet. A free-swimming tadpole stage is absent. The eggs, usually laid in clutches of ten or fewer, can be deposited in a variety of habitats, ranging from damp soil to tank bromeliads clinging to the branches of understory trees. Adults frequent similar moist habitats, climbing vegetation just like true tree frogs. They feast on mites, spiders, and small insects, especially ants. But while hunting, they must be wary of snakes and lizards, some of which relish them as their preferred menu.[170] Perhaps that explains why so many of these frogs are nocturnal, although at night they have yet another enemy that considers them a delicacy to contend with, namely bats.[171] This entombed leptodactylid must have recently hatched out of its egg (158). A puncture mark on the side of its head may explain how it ended up in resin. Whether the wound was inflicted by a beak or a fang, it indicates that our small amphibian was being assaulted by a predator when it

158. Frogs of the genus *Eleutherodactylus* (Leptodactylidae) are the only adult frogs to have been captured in amber thus far (C).

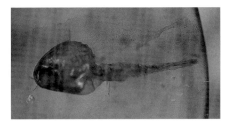

159. A tadpole in amber? Tadpoles can develop in tank bromeliads and one could have fallen into resin on the trunk of a supporting algarrobo tree.

leaped or was dropped into the sticky deposit. The mucous secretions on portions of the body may have been a defensive response.

In this small nugget is what appears to be a tadpole (159). This quite rare find could be explained if the tadpole had been developing in a tank bromeliad on the algarrobo.[172] Other types of frogs rear their young in these micropools (as mentioned in the section on aquatic life), where the mother frogs return to deposit sterile eggs as food for their tadpoles. Next to this tadpole was an egg that could have been placed in the water as an extra food supply.

Two types of fossil lizards have been discovered, and representatives of both—the anoles[173] and the geckos,[174]—still can be found in the Dominican Republic. The slender anoles, well known for their chameleonic ability to change skin coloration depending on the background, feed on arthropods inhabiting tree trunks and forest litter. Many species rest on their own personal perch, keeping an eye open for signs of prey. When a possible candidate reveals itself, the anole rushes from its lookout, catches and devours the quarry (if successful), and then returns to its observation point to resume its vigil. Insects and spiders comprise most of their needs, but frogs and smaller lizards are also on the menu. While surveying his surroundings for food, a male anole may scurry off to challenge another male that has entered his territory and, assuming a defensive posture with the fold of skin (dewlap) under his neck extended, bob his head and body up and down. If such scare tactics fail to intimidate the intruder, then a fight may ensue.[175]

157B

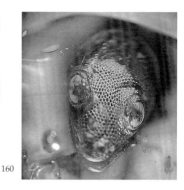

160

157B. The largest lizard found in any amber deposit, this *Anolis,* apparently a young male that was probably surveying his territory, may have been attacked by a predator and knocked into the resin (C).

160. While the other parts of this gecko (Gekkonidae) are missing, the head remains as a hunter's trophy (C).

Most of the anoles captured are juveniles who were probably either pursuing or being pursued when they landed in the resin. Desperate anoles, chased up trees and out onto the tips of the branches by snakes, will hurl themselves off the limb, plummeting to the ground, and then scurrying away as fast as possible. Many snakes follow the same tactics, however, so such aerial feats may not save the day because the snake can recover from a fall in almost the same amount of time as the lizard.[176] As if things were not bad enough, some monkeys make a repast of anoles, and young lizards may even serve as dinner for large praying mantids. A rare find is a large three and a half inch anole, the largest lizard ever to be discovered in any amber deposit (157B). Evidence of a dewlap indicates that this is a male that was probably surveying his territory when attacked by a predator.

Geckos are more common than anoles. While many geckos were ground dwelling, the adhesive pads on their toes made them excellent climbers, even on the smoothest of surfaces. The fossil geckos were probably ensnared in the resin as they shimmied up and down tree trunks. The head of this gecko with its uncanny staring eyes is remarkably preserved (160). Most

161. An entire gecko such as this
Sphaerodactylus (Gekkonidae) is rare.
Could the notch on the leaf have been
made by an insect the gecko attempted
to seize? (C).

geckos do not have movable eyelids and thus cannot blink or
close their eyes. In spite of their fixed gaze, their eyes are covered with a transparent skin, which they cleanse periodically
with their tongue. They vocalize with noises that sound like
croaks and squeaks. Such reverberations may have echoed
though the still evening air of the amber forest. At night, many
relinquish their hiding places under bark, in logs, or debris to
emerge and forage when fewer predators are about.

Geckos are largely insectivorous, suggesting how they got
caught in resin. Resembling a painting, the image of the
Sphaerodactylus could be entitled "Gecko with leaf" (161). Due
to the feeding damage on the leaf, one may assume that a de-
lectable insect browsing on the plant was being watched intently. Suddenly the hungry lizard lunged, propelling itself as
well as the leaf into a patch of resin. There is no way of con-
firming this past scenario unless perhaps the gecko still cradles
the savory insect inside its throat.

The presence of snakes in the primeval forest is represented
by at least two remnants of shed skins of these reptiles in amber. One (162) has been tentatively identified as that of a blind
snake, representatives of which still occur on all of the Greater
Antilles today, including seven endemic species on Hispaniola.

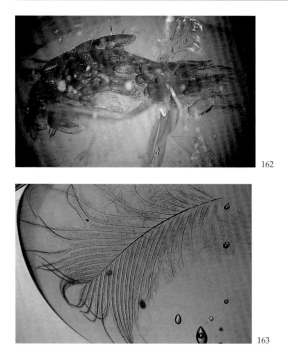

162. Portion of a snake skin, possibly that of a blind snake (Typhlophidae: Serpentes) that pushed against the bark of an algarrobo tree.

163. Although a variety of feathers occur in amber, this is the only one that has been identified, belonging to a piculet of the endemic genus *Nesocites* (Picidae) in the woodpecker family (C).

These primitive snakes possess blunt heads, diminutive eyes, and shiny scales, and they burrow in rotting logs and soil in search of termites, ants, and other invertebrates.

While entire specimens of frogs and lizards have been unearthed, only remnants of birds and mammals have been detected, making their precise identification much more difficult. Fragments of small downy feathers and portions of larger plumes suggest a diversity of bird life in that pristine environment. Only one feather has thus far been identified to a particular group of birds (163). Scientists have determined that it belongs to a small bird in the woodpecker family called a "piculet," and the feather is remarkably similar to those of the

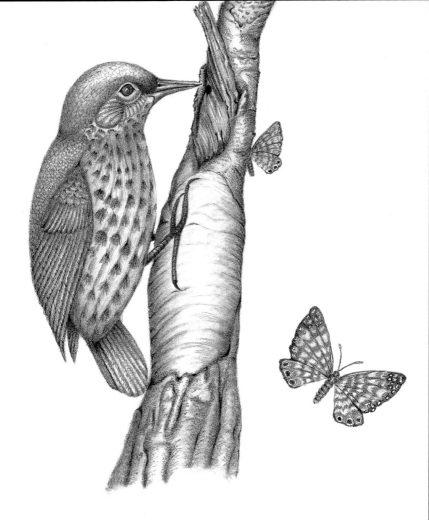

A piculet similar to this one lost a feather (photo 163) while searching for insects on the bark of the algarrobo tree. In the same area occurred riodinid butterflies (53A), which visited flowers, mated, and searched for sites to deposit their eggs.

164. This tuft of rodent hair could be that of an ancient Hutia similar to those of the genus *Plagiodontia* (Capromyidae). Hair identification can be determined not only by microscopic features but also by associated parasites (C).

Antillian piculet that represents an ancient lineage still occurring in the Dominican Republic today.[177] Smaller than common woodpeckers, they lack the hard, stiff tails most woodpeckers use to prop themselves up against the trunk of a tree. In fact, piculets resemble nuthatches and, like them, creep along branches looking for food. Their loud repeating calls must have echoed in the early morning throughout the mist-shrouded forest. The Antillean form, and probably the fossil, nests in tree holes and rotting trunks, feasting on ants, termites, and wood-boring beetles, and their habitat includes humid as well as partially dry forests.

The most prevalent traces of mammals are hairs. In most cases, only a few strands remain; however, in exceptional cases as illustrated here (164), complete tufts or clusters are preserved. Sometimes, epithelial cells are retained along with the hairs, and occasionally some minuscule ectoparasites are also embalmed.

Fortunately, all hairs have microscopic features that, if clearly visible, can identify the owner as belonging to a particular order, family, or even genus of mammals. The hair strands shown here (164) have been identified as belonging to a rodent.[178] Identification was further facilitated by two minute ectoparasites lodged among the strands, a fur mite and a fur beetle. Modern-day representatives of these ectoparasites fasten themselves to the coat of rodents. A likely candidate is the Hutia, or Coney. Hutias, endemic today to the Caribbean area, have largely become extinct in the past one thousand years as a result of habitat destruction, hunting and introduced predators. Living in dense vegetation or what is left of forests in the Antilles, a few of these secretive rodents are maintaining their populations only on the most remote islands where predators are absent. Only a single species remains in the Dominican Republic today.[179]

Hutias resemble muskrats in size and in possessing a tail which is almost free of hair. They are agile climbers, a skill which has probably aided their survival to the present.[180] However, in Cuba, arboreal habits alone do not safeguard them from becoming a meal for jumping crocodiles. The Cuban Hutia frequents swamps, where it collects food from a variety of wetland trees. These rodents incur the risk of being devoured by freshwater crocodiles as they swim from tree to tree. While perching high up in a tree would appear to offer immunity, scaling a small tree or shrub is no protection against these crocodiles, which have the astonishing ability to jump twelve feet out of the water and snatch defenseless Hutias from their tenuous holds among the branches. Apparently the crocodiles depend heavily on them as a major food item.[181] It is not known but possible that the range of these crocodiles extended to Hispaniola back at the time of the amber forest.

Portions of several mammal bones were recently detected in amber. From remnants of thoracic vertebrae and rib portions, researchers concluded that the source was a small insectivore belonging to the now extinct genus *Nesophontes*.[182] These small mammals were eradicated quite recently (within the past one hundred years) as a result of human activity. Not much is known about their habits other than that they were about the size of a shrew and fed on a range of insects. Their demise was probably a result of habitat destruction and the introduction of cats and dogs, against which *Nesophontes* had little defense.

As already discussed, one type of indirect evidence of the vertebrates populating the prehistoric forest is the presence of fossil parasites that depended on them for survival. Aside from

A now-extinct insectivore (*Nesophontes*) searches for insects, while above on a branch two predators, a preying mantis (photo 136) and a gecko (161), both eye the same bush cricket (50) as a potential meal.

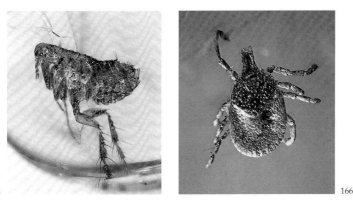

165

166

165. This male flea of the genus *Rhopalopsyllus* (Rhopalopsyllidae: Siphonaptera) probably tormented a rodent (C).

166. A male hard tick of the genus *Ambylomma* (Ixodidae: Acari) provides indirect evidence of vertebrates in the amber forest (C).

fur mites and beetles, more traditional ectoparasites have been preserved. Such ectoparasites comprise various arachnids and insects that usually sojourn on the body of assorted vertebrates, usually just long enough to engorge themselves with the victim's blood. One typical ectoparasite is the flea (165).[183] Restricted to birds and mammals, each species of flea has a specific host or group of animals upon which it feeds. This particular individual belongs to a family that today resides mainly in the New World tropics, where they torment various mammals, especially rodents.

Ticks found preserved fall into two morphological categories, hard (ixodid) and soft (argasid). Hard ticks usually have mouthparts protruding from the front of the head, while soft ticks have mouthparts concealed beneath their bodies. Both types of ticks were abundant in the amber forest. The hard ticks don't reveal much about their hosts other than they were vertebrates. This type of hard tick (166) is very catholic in its tastes and will attack lizards, birds, and mammals.[184] Soft ticks, however are more selective and occur only on birds and mammals. Adjacent to the one featured here (167) is a scat belonging to a rodent.[185] Soft ticks in Puerto Rico today feed on rats, and it is

167. Most soft ticks like this male *Ornithodoros antiquus* (Argasidae: Acari) occur on mammals; the adjacent fecal pellet suggests that the host was a rodent (C).

possible that this fossil was feeding on a rodent long ago. Based on the size of the fecal pellet (25 mm long by 8 mm wide) relative to the size of a Hutia, it is credible that a Hutia was the carrier of this arachnid. Ticks transmit a number of disease-causing pathogens to their hosts and perhaps eventually we can determine if this was the situation millions of years ago.

From the range of sizes and microscopic features on mammalian hairs, one can infer that more than just rodents and insectivores roamed the ancient tropical forest. Did carnivores wander about the island of Hispaniola? Thus far one source of hair was thought to belong to a carnivore, and it was attached to the bamboo seed discussed earlier.[186] There are accounts of jaguars carrying such bamboo seeds entangled on their pelts today, and we suggest that an ancestor of these predators may have dwelled in that earlier jungle.

Other products of mammals were their scats. Already mentioned was a rodent scat that quite possibly originated from one of the extinct species of Hutias. Other fecal matter of various sizes and shapes hint at the presence of a variety of small mammals in the amber forest. Mammalian dung is in fact a highly desirable resource in tropical forests, and a number of insects compete for such prizes.[187] Attracted by smell, it is first come, first served as scarab beetles and flies hone in on the

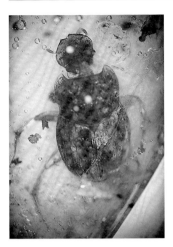

168. Dung beetles such as this *Canthidium* species (Scarabaeidae: Coleoptera) may sometimes be quite specific in their choice of mammalian refuse (C).

meal. Some dung beetles are quite choosy in their selection of this resource. Certain present-day *Canthon* beetles collect monkey dung, often from leaves, since simians defecate from the treetops. Male beetles will began working the dung into small balls while still arboreal, eventually falling with a portion of the mass to the ground, where they begin rolling their prize off to the side. The females wait patiently, sometimes even hopping aboard the rolling lumps, until the male begins to bury his prize. The fertility ball is eventually entombed with the egg enclosed. This action must be quick: if dung-hunting flies locate and lay their eggs on the mass, there won't be enough nourishment left for the beetle larvae. From their fossil remains it is apparent that a host of dung beetles occurred in the ancient forest, many possibly specializing in certain types of dung. Some may have perched on leaves, as do their present-day descendants, waiting for the right sound or smell; the species of *Canthidium* shown here (168) may have become entrapped in resin while doing just that. If our knowledge of the dung preference of present-day beetles were more complete, these fossil insects could provide valuable information about the types of mammals that existed at the time of the amber forest.

Hoards of insects, including *Culex* mosquitoes (169), biting midges, sand flies (170), and horse flies, all of which required

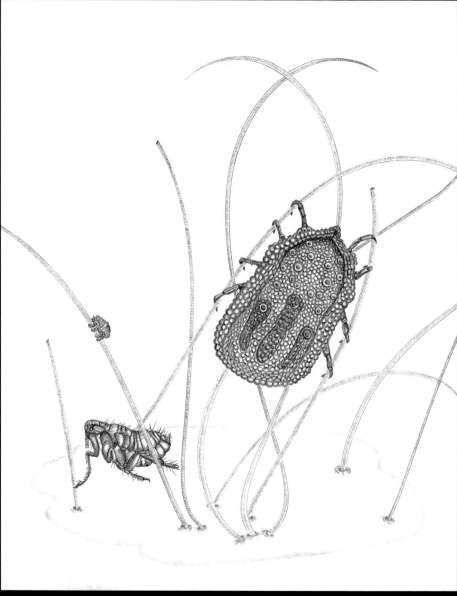

Depicted here are three mammalian parasites that could have victimized the same rodent in the amber forest. A large soft tick (photo 167) climbs up the hair of its host after taking a blood meal, while a flea (165) remains close to the skin of its victim. A tiny listrophorid fur mite tightly clasps a single strand of the rodent's hair.

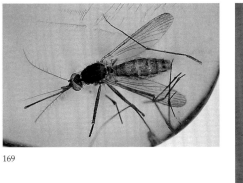

169

170

169. One of the most common genera of mosquitoes in amber is *Culex* (Culicidae: Diptera), which could have developed in standing water between bromeliad leaves (C).

170. Sand flies like this female of the genus *Lutzomyia* (Phlebotominae: Psychodidae: Diptera) were another group of bloodsuckers and may have transmitted the protozoan *Leishmania* to vertebrates in the amber forest (C).

the blood of a vertebrate for survival, must have plagued the animals of the primordial forest. Throngs of mosquitoes hovered throughout the forest layers and, as mentioned earlier, many probably depended on the reservoirs in tank bromeliads for breeding. Perhaps sometime in the future, by examining their intestinal contents or salivary glands, it may be possible to determine if these biting flies were in fact harboring various pathogens. Mosquitoes transmit viruses, protozoa, and nematodes, while biting midges, which probably took blood from invertebrates as well as vertebrates, are capable of transmitting virus pathogens to vertebrates. Sand flies or phlebotomines also transmit protozoa that cause cutaneous leishmaniasis in the Neotropics.[188] Who knows what these ancient bloodsuckers were transmitting to their victims.

Other flies in this ancient ecosystem had intimidating rostrums as long as mosquitoes. Surprisingly, such beaks were not used to take blood but were adapted to obtain nectar from tubular-shaped flowers. The adults are known to land on

171. While possessing the beak of a mosquito and appearing ready to feast on blood, this female cranefly of the genus *Geranomyia* (Tipulidae: Diptera) only uses its rostrum to imbibe nectar from flowers, especially those of the Compositae (C)

flowers, especially in the evening after the bees have departed, and insert their beaks into the petals in a manner reminiscent of hummingbirds. One of these insects was the beaked cranefly which is depicted here (171). Members of this genus are especially attracted to the small florets of composite flowers.

3

Reconstruction of the Amber Forest

THERE ARE many types of tropical forests today, and the Holdridge classification of world plant formation is used here to characterize the amber forest.[1] The important features in this classification are temperature, rainfall, latitude, and altitude. This set of parameters determines which plants will grow and diversify in an area, and the plants in turn will decide the types of invertebrates and vertebrates that will reside in or visit the forest.

In this book, an assortment of plants and animals from the algarrobo forest have been presented. This sampling of the flora and fauna provides us with enough evidence to piece together the type of forest that existed some 15–45 million years ago. Clues derived from this assemblage also suggest reasons for the extinction of many lineages from the Greater Antilles. It is clear from the foregoing discussions that many of the animals and plants that existed in the distant past have no immediate descendants in the Dominican Republic today nor in fact in the entire Greater Antilles, suggesting that the factors responsible for the elimination of these lineages did not influence just the island of Hispaniola. Thus the biodiversity of life back then appears to be greater than what exists on the same landmass today.

Appendix B lists families and genera of biota that have been found in Dominican amber, representing the largest fossil assemblage of terrestrial invertebrates in a tropical environment.

More will be added to this sizable number as scientists continue to examine these fossils. Although only a fraction of them could be discussed here, the knowledge gained from all of those listed has been incorporated into the reconstruction of the original forest.

How did the entombed plants and animals reach the island of Hispaniola? Probably by several routes. Certainly the ancestors of many migrated onto the Proto-Greater Antillean archipelago when it was still in contact with North and South America, much like Central America today. According to this scenario, known as the "vicariance theory," the Proto-Greater Antilles consisted of a huge interconnected floating garden and menagerie carrying animals and plants from both North and South America into the Caribbean Sea. As this floating ark moved eastward, it bumped up against ridges extending out from the North American land mass. With a lowering of the sea level, temporary land bridges of some years' duration may have appeared later to allow for the arrival of other biota.

If such land bridges did form and remained for substantial periods, then ungulates, marsupials, and carnivores should have reached the island.[2] A recent discovery of a primitive fossil rhinoceros dating from the Eocene in Jamaica proves that ungulates did indeed reach the Proto-Greater Antilles.[3] This find suggests that there are probably many other surprises still awaiting discovery regarding the types of mammals that resided in Hispaniola at the time of the amber forest.

Overwater dispersal was another way groups could have reached the isolated land masses. Some forms, including insects, bats, and birds, could have made landfall on the islands as they were being carried to their present-day location. Floating logs and masses of vegetation may have rafted insects as well as small terrestrial vertebrates such as lizards and snakes. Air currents and storms could possibly account for the presence of some insects, spiders, and mites. Although the prevailing air currents are from the northeast,[4] they might have aided in the movement of insects from island to island. Of course, just arriving on an island does not mean that colonization will materialize. Unless a pregnant or parthenogenic female arrives, more than one individual of the same species must touch down at the same locality at the same time. Then the new arrivals must find an unoccupied niche or compete with the already established residents in a new niche, which would not be an easy task. In the case of phytophagous forms, their survival would depend on the presence of certain types of vegetation.

From ongoing studies, it appears that most of the ancient flora and fauna probably reached Hispaniola some 40–60 million years ago while it was still in immediate or close contact to the mainlands of North and South America. Other biota that arrived via temporary land bridges, drifting flotage, air currents, or under their own power certainly added new elements to the ecosystem, but such arrivals were probably sporadic and fortuitous rather than continuous.

While amber inclusions present a rich and varied view of many invertebrates that inhabited the ancient forest, they are obviously deficient when it comes to the larger vertebrates that must have occurred at that time. Their presence can be determined in some part both by fossils preserved by other means and by an assessment of the extant species that by their morphology and distribution can be considered as ancient lineages. The next few pages will be given to a discussion of other vertebrates, which possibly lived in the ancient forest, but for which no direct fossil evidence from the amber forest currently exists.

In dealing with mammals, we have already presented evidence in amber for a small insectivore and rodent, but there is a larger and endangered endemic insectivore on Hispaniola which could have existed during the period of amber formation. This is the solenodon, a quite bizarre-looking animal with a greatly extended flexible snout used to root in rotting wood and soil for invertebrates.[5] These creatures also dine on frogs, lizards, and young birds. Their large claws are used to pin down the game, while the lower jaw, equipped with large protruding incisors, scoops up the flesh. Their toxic saliva immobilizes the prey before it is eaten. Solenodons have been described as slow, dull creatures that shuffle through the forest on flat feet. They live in burrows, rock dens, and hollow logs, coming out at night to forage. Although limited today to Hispaniola and Cuba, there were many more solenodon species in the past, all extinct now as a result of habitat destruction and the importation of dogs and cats. It is likely that within twenty years, the two existing species will also vanish. The progenitor of these insectivores may have entered the Proto-Greater Antillean landmass while it was still connected to North America.

Another group of mammals that may have entered Hispaniola in the same manner were the sloths. At least ten genera of West Indian fossil sloths are known, and at least six endemic species occurred in Hispaniola. Paleontologiocal evidence established that these megalonychid sloths arrived on the Greater Antilles over 30 million years ago, with the oldest fossil sloth found in Miocene deposits in Cuba.[6] It is possible that preserved sloth hairs will be found in amber someday, since a number of the extinct Hispaniolan sloths apparently were arboreal while others were ground dwellers. These sloths were the size of dogs, not huge like some of the North American forms. Surprisingly those of Hispaniola appear to have survived the longest of all the ground sloths, quite possibly overlapping with humans.[7] A variety of plants undoubtedly represented the major portion of their diet, although insects and possibly dead or dying animals may have also been eaten.

As mentioned in chapter 2, a tuft of hair identified on the basis of its structure and associated ectoparasites may have come from a Hutia.[8] These endemic rodents of the family Capromyidae had become quite abundant in Hispaniola by the Pleistocene, as determined by fossil bones in caves and kitchen middens.[9] Today most of the genera have become extinct, and these short-legged rodents are represented by only a single or possibly two extant forms. These animals, consisting of both ground and tree dwellers, are thought to have entered the West Indies from South America at least by the early Miocene, which means they could well have existed in the prehistoric forest.[10]

Were monkeys swinging through the canopy and understory? Did the woods ring with their yipping whines, clucking alarms, squeals, and chatter? Did the sound of dropping vegetation, the hammering of fruits against tree trunks, and thrashing branches echo through the stillness of the morning hours? It's possible! Early reports of monkey remains in Cuba, Jamaica, and Hispaniola exist[11] and more recently monkey bones have been recovered from caves in Hispaniola.[12] Scientists are debating whether these remains came from endemic primates whose ancestors had survived for millions of years in the West Indies, or if they just represent pets carried to the island by Amerindians. However, at this time it appears that the Hispaniolan remains predate human occupancy. It has been suggested that the monkeys were tamarinds, while some claim they were squirrel monkeys (*Saimiri*) and still others argue that they were capuchins (*Cebus*), or organ-grinder monkeys. At least the fossils establish that one primate probably survived for a considerable amount of time on the island of Hispaniola—but if it was 20 million years ago is a question that must await further investigation. Monkeys could have reached Hispaniola by island hopping or migrating along land bridges at periods of low sea levels, especially if the island came close to the mainland as it slowly drifted eastward into the Caribbean. One author claims that the Tainos hunted extinct squirrel monkeys in the Dominican Republic.[13]

Most certainly there were bats in the forest, representing a variety of sizes and food preferences. In fact, some hair strands in amber closely resemble bat hair, including a few associated with a nymphalid caterpillar (53A). These flying mammals were established in the Americas some 50 million years ago, and their powers of flight surely enabled them to reach Hispaniola when the resiniferous forest was thriving. One genus, endemic only to the Greater Antilles today, is a nectar-feeding bat, *Monophyllus*.[14] Small creatures, they have elongated muzzles and a narrow tongue that is papillate at the tip. The tongue can be extended far into the throat of a flower to reach the nectar, and while this is happening pollen is brushed off its head. These animals are predominately nocturnal, which explains why nectar-feeding bats can be important pollinators of night-blooming plants, especially those that attract night fliers with large, whitish flowers.

Aside from the nectar feeders, there were presumably also fruit bats (Phyllostomidae). These are acrobats specializing in picking fruits from trees and conveying them to a convenient roost where the feast begins. During this meal, indigestible portions such as fibers and larger seeds are cast on the ground, and this habit makes such fruit eaters important seed dispersers in tropical forests today.

At least two groups of insectivorous bats are endemic to the Greater Antilles, and their ancestors may have winged through the amber forest. Big brown bats (*Eptesicus*) roost in hollow trees and caves and cruise over open areas in search of flying insects. Tiny in comparison with these are the funnel-eared bats (*Natalis*), known for their habits of roosting in masses, sometimes in the thousands, in damp caves. Only 2–3 inches in length, these miniatures retain a wide band of crinkled skin on their wings, which probably accounts for their fluttering flight as they pursue insects.

Up until quite recently, there had been no evidence of endemic ungulates, either extinct or extant, in the West Indies, and the dictum was that no ungulates ever reached the Greater Antilles. However, scientists have now discovered a primitive rhinoceros (rhinocerotoid) known as *Hyrachus* that roamed the early forests of Jamaica some 50 million years ago.[15] These early ungulates became extinct when Jamaica became inundated some 40 million years ago. However, it is possible that *Hyrachus* or a closely related genus reached Hispaniola, and since this large island was never completely submerged, these or related ungulates could have survived in the amber forest. *Hyra-*

chus was moderately sized and probably frequented both forested and swampy areas, with habits similar to those of the modern tapir. This interesting discovery is highly significant in establishing a new group of mammals in the history of the Greater Antilles, but it is also important in showing that the fossil record of any geographical area is very incomplete and that it is not possible to state that a group never existed simply because fossils have not been found.

The carnivores are another unknown. They would be expected in this type of ecosystem, yet unequivocally ancient fossils have not yet been recovered. There were felids and highly specialized saber tooths in North America at least 30 million years ago,[16] and it is feasible that some reached the Greater Antilles. Some hairs tentatively identified as belonging to a carnivore were associated with a bamboo seed in amber[17] but there are no other records of endemic carnivores in Hispaniola. Some canid bones have been found in Cuba, and while some scientists have described these as extinct genera of wild dogs (*Cubacyon* and *Paracyon*), others feel they came from pets introduced by the Amerindians.[18] The small ground sloths and Hutias would have been ideal prey for medium-sized wild dogs or cats.

Snakes were probably fairly abundant at that time as well. Aside from blind snakes, whose possible remains appear in amber, there are boa remains from the Miocene of Puerto Rico[19] and boas (*Epicrates*) naturally occur on Hispaniola, suggesting that they could have extended back in time, quite possibly some 20 million years.[20]

Did crocodiles navigate the rivers coursing through the tropical forest? These giant reptiles can be found in Hispaniola today, but Cuba is the only island in the Greater Antilles which still has a truly native crocodile, commonly known as the "jumping croc" because of its habit of snatching Huitas from the branches of swamp trees.[21]

Although fossils are lacking in Hispaniola, an endemic freshwater turtle (*Chrysemys decorata*), restricted today to lakes in the southeastern portion of the Dominican Republic, may have had ancestors that thrived in the ancient sylvan ponds.[22]

Certainly a variety of birds filled the air back then, as evidenced by the range of feathers found in amber. Some of these feathers could have fallen from hummingbirds similar to the native Hispaniolan emerald hummingbird. These tiny gems, slightly larger than the ruby-throated hummingbird, are smothered in various hues of radiant green, with males displaying a velvety black breast patch.[23] Even more dazzling would be the tanagers, two of which are restricted to Hispaniola today. The black-crowned palm tanager, with its colorful plumage of white, black, gray, and yellow-green resides today in lowland woodlands and thickets. Chances are that delightful todies also flew through the algarrobo forest, since these unique kingfisher-like birds are today solely confined to the Greater Antilles. The narrow-and broad-billed todies of Hispaniola, both the size of hummingbirds, display quite breathtaking opalescent green-black body colors highlighted with deep-red throat patches and rosy sides. Occurring in dense humid forests, they feast on insects caught in the air by darting out from their exposed perches and then returning with the prize.

Could there have been brightly feathered, gaudy parrots? Macaws recorded in historic times in Cuba became extinct in the nineteenth century, leaving behind a few skins in museum collections as evidence of their existence.[24] One group of parrots, the *Amazona,* are only found in the West Indies today, with each island harboring its own unique species. These flamboyant birds probably arrived in the West Indies at an early date and then diversified on the various islands. The Hispaniolan emerald parrot dines chiefly on fruit and nests in tree cavities or rock crevices. Hispaniola also has its own unique parakeet, the *Aratinga,* which is found in the mountainous areas of the island.

Just as colorful as the parrots and parakeets, with its resplendent plumage and long tail feathers, is the Hispaniolan trogan, which feasts on both insects and fruits. No one knows when the ancestors of these birds arrived in Hispaniola, but they could have existed in the primeval forest.

The Greater Antilles was also at one time home to some of the largest birds of prey. The remains of immense eagles, fleet hawks, and rapacious falcons from Cuba, Hispaniola, and the Bahamas indicate that these predators probably hunted the land mammals that once colonized the islands.[25] A colossal barn owl that at one time surveyed the forests of Cuba and Hispaniola was large enough to capture young sloths, although rodents were probably among the main choices of these giants. To scavenge the remains of pillaged sloths and rodents were huge condors, whose bones have been found in Cuba. All of these birds are now extinct, but their ancestors may date back to the time of the old forest.

Shore birds probably frequented the banks of waterways that meandered through the forest.[26] Rails, similar to the almost flightless species that congregates in the Zapata swamp of Cuba today, could have been one of these ancient bird lines. Another is the native West Indian tree duck, which displays behavior unusual for ducks. These nocturnal, goose-legged birds perch and even nest high in trees, sometimes rearing their young on epiphytic bromeliads. This behavior would appear to be expedient for avoiding terrestrial predators, which once might have included still unknown forms besides the possible solenodons. Clearly there was a host of other birds, perhaps to be identified someday from the feathers that occur in Dominican amber. The piculet that we featured here is just the prelude.

Other Tertiary fossils from Hispaniola of the same time period as the ancient forest include a Miocene freshwater cichlid fish.[27] These tropical fish, which are quite colorful and highly prized by aquarium enthusiasts, can reach up to a foot in length and probably feasted on some of the aquatic insects discussed earlier. Many of these native fish are extinct or near extinction today, partly as a result of the introduction of the aggressive freshwater perch, *Tilapia*, from Africa.[28]

Fossil bones of manatees in Cuba dating also from that period verify that the ancestors of the now endangered West Indian Manatee probably were quite abundant in the blue waters surrounding Hispaniola at the time the amber forest was thriving.[29]

Before proceeding further, we must point out that two important restrictions must be considered when undertaking the reconstruction of the amber forest—time and space. Necessity compels us to represent all the fossils as occurring within the same general forest during a continuous time period. This indeed would probably be the case if we can assume that Dominican amber was formed 15–20 million years ago, as has been recently concluded by scientists examining fossil protozoa in the surrounding rock matrix,[30] and that all of the amber from both the northern and eastern portions of the country was produced by one extensive forest during this time period. However, amber pieces from separate mines exhibit physical and chemical differences which could signify that they differ in age or, alternatively, that the resin was subjected to different tectonic conditions after burial. In fact, another researcher has suggested that at least one mine contains amber up to 45 million years old.[31] Could the fossils represent a series of different ages and display extinct life from more than a single forest? The second restraint deals with the homogeneity of the forest. Since the amber mines are scattered across the country today (but roughly within the same latitude), would altitude, climatic differences generated by prevailing winds, or soil differences have created more than a single ecosystem? Whether all the ancient biota flourished in the same kind of forest within a certain time period, or whether a portion of the fossils date from the Eocene in one type of forest and another are representative of a later Miocene forest, is a debatable point. However, since there is no study yet indicating that speciation of lineages occurred in amber biota, we conclude that all of the fossils occurred in a similar type of forest, albeit fairly extensive, that existed for at least several million years, possibly longer. It is probable that an established algarrobo forest experienced some catastrophic event that caused the trees to exude copious amounts of resin for some time, thus representing the Dominican amber deposits. Further analysis of the fossils and geological events in the Caribbean during the Tertiary might help to resolve these questions.

Judging from the plant and animal fossils, there is no doubt that the forest was tropical. This is corroborated by the geographical position of the Dominican Republic in the past as well as the present. However, evidence, which will be presented later, suggests this tropical forest experienced climatic fluctuations at some point, changing its biota forever.

The principle of behavioral fixity has been applied to characterize the fossil flora and fauna in Dominican amber.[32] This well-established principle proposes that the behavior of fossils will be comparable to the behavior of their present-day descendants at the generic and in some cases at the family level. Thus, by comparing fossil biota with their closest living descendants at the generic or family level, it is possible to obtain relevant information about the likely behavior and preferences of these extinct species.

Today, tropical forests are categorized as very dry, dry, moist (humid), wet, and rain forest, depending on the average total annual rainfall.[33] Inversely, plant distribution can be employed to indicate the forest designation since many plant genera are restricted to one or two forest habitats. Analyses of the plant genera cited here and listed in appendix B clearly illustrate that all plants are typical of tropical moist forests today. No other forest biome contains so many of these plants, especially *Acacia* and *Mimosa,* which are represented in amber by several species. Furthermore, algarrobo trees today, while occurring in a variety of habitats, are most abundant in tropical moist and dry forests.[34] It should be noted that geographical areas often consist of more than one forest regime, depending on size and topography. For example, the portion of a forest facing the prevailing winds may have a tropical wet climate, while on the other side of the slope there may be conditions of a tropical moist forest, and further inland the climate may dictate a tropical dry forest. In the case of this ancient forest, nearly all of the biota listed in appendix B that have descendants at the generic level would be typical of animals and plants found in a Neotropical moist forest today. In fact, the plants that still have generic representatives in the Dominican Republic today all occur in remnants of moist tropical forests, especially in the Reserva Cientifica Natural Isabel de Torres, an isolated park surrounded by sugar-cane plantations in an area south of Puerto Plata on the northern coast of the Dominican Republic.[35] In addition, with the exception of Thymeliaceae, *Roystonea*, *Swietenia*, and *Peritassa*, all of the plants found in amber have modern descendants in Barro Colorado Island,[36] which is classified as a tropical moist forest.[37]

A particularly comprehensive study of entombed insects surveyed the darkling beetles, Tenebrionidae.[38] Of the sixteen genera of beetles examined, practically all belonged to taxa whose contemporary descendants inhabit moist forest biotypes; only one genus survives today in arid or subarid habitats. The moist forest borders the dry forest biome, and movement from one to the other could be possible. But within each biome there also exists a considerable diversity of habitats. Within the moist forest type, both evergreen and semideciduous vegetation occurs. Of the New World generic descendants of insects reported from Dominican amber, more can probably be found in Neotropical moist forests than any other forest type today.

Contemporary tropical moist forests scattered throughout Central and South America are distinguished by a dry season of 3–4 months succeeded by a wet interval of 8–9 months. The presence of gemmae on mosses and liverworts and the occurrence of tank bromeliads suggest a dry cycle prevailed back then as well.

Additional evidence of a dry season is difficult to obtain with most invertebrates since when the weather changes they can move from one biome to another in search of food and shelter. However, a dry period is suggested by the presence of a replete

A fragment of the amber forest, as reconstructed from fossils in Dominican amber. The emergent tree is the caoba tree (*Swietenia*) reaching 135 feet, the canopy layer is depicted by two algarrobo (*Hymenaea protera*) and a nazareno (*Peltogyne*), the subcanopy and understory by a *Roystonea* palm and fig (*Ficus*), the shrub layer by various palms and *Acacia,* and the forest floor by pega-pega (*Pharus*) and others. Epiphytes include orchids, bromeliads, and ferns. Vines and lianas are stretched between and among the vegetation.

worker *Leptomyrmex* ant. These acclimatized ants customarily
store liquids in their body only during prevalent dry periods.
Harvester ants too are typically found in biomes with a notable
dry spell, which reduces the germinating potential of their collected seeds.[39]

Tropical moist forests contain several strata or layers of vegetation.[40] Lofty canopy trees towering from 80 to 130 feet
would equate to the algarrobo, cativo, nazareno, and sebo, with
the caoba rising up as an emergent. A subcanopy layer of trees
attaining 40–80 feet would incorporate the sigua. The understory layer with vegetation ranging in height from 20 to 40 feet
would encompass palms, *Acacia*, *Mimosa*, and *Trichilia*. The
shrub layer of 5–20 feet would contain palms and other *Acacias*, while the forest floor would be relatively bare with pega-
pega, ferns, and ground cover in the form of moss and liverworts, which are also epiphytic in habit. Abundant herbaceous
and woody vines and vibrant epiphytes drape the trees. The
flora of the extinct forest clearly conforms to a tropical moist
forest.

Today's tropical moist forests embrace a prodigious abundance and diversity of species on the face of the earth, comprising over three-quarters of the planet's vegetation.[41]
Mankind has perilously reduced these crucial forests to one-
third of their original distribution. With the cutting of every
hectare of moist tropical forest, we irrevocably lose potentially
valuable plants that could be used in agriculture or pharma-
cology, not to mention a multitude of invertebrates and vertebrates whose lifestyles are intimately interwoven with those
plants.

The tropical moist forest that existed millions of years ago
was unparalleled—nowhere is there an extant equivalent.
There are no forests with *Mastotermes* termites, *Leptomyrmex*
ants, or *Halovelvia* marine bugs in the Americas today. Aside
from the fact that probably all of the species, a number of the
genera, and at least one subfamily are now extinct, many lineages are completely gone from Hispaniola today. It is not

surprising that the majority of the fossil fauna and flora display affinities with Central and South American lineages since many of the life forms originated from those areas. However, it is clear that the composition of the fauna and flora in the amber forest was different from that of the present day.

Whether any tropical moist forest persists in the Dominican Republic is questionable. Some authorities believe that relict patches still survive here and there, but others doubt any vestiges of endemic tropical forest habitat remain.[42] This book has made it obvious that many genera and species of animals and plants in Dominican amber are now extinct from Hispaniola. When and how did this unique flora and fauna disappear? Did the indigenous biota suffer from the influence of man through habitat destruction and competition with introduced exotic organisms?

Tropical bees can be used to illustrate the pattern of extinction that probably beset these fossil lineages. All of the seven genera or subgenera of bees found in amber are now absent from Hispaniola, while their relatives persist in Central and South America.[43] Could these bees have been eliminated by habitat destruction or by the introduction of the European honeybee into the West Indies in the seventeenth century? Evidence implies that in reality the native bees had already vanished before the Europeans arrived in Hispaniola. An early report from 1674 confirms that at the time the honeybee was disseminated, noticeable bees were absent.[44] Certainly the stingless bees would have been observed if they had persisted to that period.

In addition, evidence of the types of insects in the region some 200–300 years ago can be acquired from an examination of copal from the region of Cotui. This subfossilized resin is informative because of what it does not contain. Stingless bees are particularly abundant in amber because they habitually utilized the resin deposits as a resource for nest construction. However, in Cotui copal, which is quite fossiliferous, no bees of any type have been reported, nor does it contain any of the other insects that characterize those early deposits. This further substantiates that the bees had perished before the European colonists arrived.

Could some disease have been liable for the demise of the amber bees? While there are pathogens and parasites that can annihilate honeybee colonies today, the diseases of tropical bees are poorly studied.[45] However, on the basis of what is known and even including the intensely studied honeybee pathogens, it is highly improbable that any single disease or even a collection of pathogens could have been responsible for the elimination of all of the fossil bees from Hispaniola, nor all of the other tropical invertebrates that have only been found embalmed.

It is more feasible that the cause of these extinctions was related to parameters that would affect a wide range of tropical life forms, such as climatic conditions. It has been acknowledged that global cooling during the Pliocene and Pleistocene transfigured the Northern Hemisphere and that the effect extended into the tropics.[46] A drop in temperature coinciding with these periods has been inferred both in the sea and on land based on the disappearance of stenotopic (forms able to survive only under a narrow range of ecological conditions) marine invertebrates and terrestrial vegetation from areas bordering the Caribbean. From the past distribution of plants, scientists have been able to confirm a cooling period in parts of Amazonia as well as northern South America. Average annual temperature drops of 3°–7° C have been proposed, which would have replaced the tropical moist forest in the Dominican Republic with a subtropical realm.[47] Tropical life forms characteristically are stenotopic and have not evolved mechanisms that permit their physiology to function under subtropical or temperate climates. After a shift in temperature occurred, the only solution for survival was migration to areas where the weather was similar to the original one. When it began to cool in the Dominican Republic, stenotopic tropical forms like bees were trapped by their island insularity. There may have been some "refugia" or areas in which the climatic changes were tempered, but it is obvious that these retreats were not adequate enough to save many of the biota.

In addition, since these populations had been isolated for millions of years, they might have been suffering from what is called "inbreeding depression," a condition of decreased genetic variability.[48] This rise in homozygosity could have increased the susceptibility of the bees and other life forms to climatic changes.

A fall in temperature may have been accompanied by a corresponding escalation in aridity, as is characteristic of glacial periods.[49] This decrease of available moisture would also have altered the structure of plant communities, which in turn would disturb pollinators and herbivores as well as the parasites and predators whose existence depended upon them.

The tropical bees have been used as an example of how a drop in temperature could have adversely affected their distribution and survival. Other stenotopic tropical life forms would also have been eradicated. The survivors would have been eurytopic (those able to survive under a broad range of conditions) forms and those stenotopic forms that might have discovered refugia adequate enough to maintain the essential resources required for the duration of the inclement cool period. However, on an island the size of Hispaniola, such refugia would have been scarce.

Exactly when did the cooling period occur? Examining global climates during the past 20 million years reveals that the first significant temperature drop reached Latin America and the Caribbean during the late Pliocene, or between 1–3 million years ago.[50] Subsequently there was a series of glacial-interglacial fluctuations which could have also endangered life forms, but the exact extent of chilling associated with them is not known. The tropical life forms in this primeval domain probably would have expired during the critical late Pliocene cool period. Of course, it is possible that some could have become extinct then and others during subsequent cooling periods. These questions cannot be resolved until more information about temperature fluctuations in Latin America is acquired.

Why didn't the ancestors of the vanished bees that survived on the mainland return to Hispaniola after the cooling event when the tropical climate was reestablished? Currently there are closely related populations of all of the extinct bees in Central and South America, yet, with the exception of two recently arrived orchid bees in Jamaica, none reside in the Greater Antilles today.

The answer lies in the dilemma of flying long distances over water. In general, most bees are not known for their trans-oceanic dispersal abilities, and during tempestuous storms, when wind updrafts could lift and transport them extended distances, most bees retreat to the safety of their nests.[51] Furthermore, prevailing winds during the entire year come from the northeast to the Caribbean and therefore would not be conducive to conveying insects from land masses toward the Antilles.[52] As mentioned above, two orchid bees, which are large bees well known for their vigorous flying capabilities, did manage to reach Jamaica from Central America.

The bees found in the Dominican Republic today are either eurytopic species that could have withstood the cooling period (but have not yet been found as fossils), or small bees perceived as having dispersed passively to the islands. In fact, small bees have been collected at heights of 1,000 feet over Mexico, so during rare times of wind pattern changes they may have been carried to the Antilles.[53]

A synopsis of the history of the amber forest biota then entails colonization, speciation, extinction with survival of eurytopic forms, and disappearence of stenotopic lines at the generic level. A cooling period probably eliminated the truly tropical lineages on Hispaniola during the Pliocene-Pleistocene. Their surviving relatives on the mainland, whose ancestors had originally populated the island, escaped the extinction event by migrating toward the equator or into refugial areas.

When the temperature readjusted after the cooling period, the continental species slowly drifted back into their former habitats, but few immigrated over water to the Greater Antilles and those islands remained depopulate. This pattern also gives credence to the idea that the origin of many biota reflects vicariance biogeography.

It would appear that during the temperature instability, *Mastotermes* termites, *Leptomyrmex* ants, *Halovelvia* marine bugs, as well as some others that were eliminated on Hispaniola also became extinct on the adjacent continents. Evidence of *Mastotermes* in Central America stems from their presence in Mexican amber (22–26 million years old).[54] This amber was produced by a tree closely related to the Dominican algarrobo. Consequently, Mexican fossils can be quite enlightening by revealing lineages that existed in Mexico at about the same time as the Antillean forest was flourishing. Why the above insects disappeared from the entire New World remains a mystery. Perhaps they were unable to migrate quickly enough or find their preferred habitat, or their migration south may have brought them into an unfortunate competition with more aggressive species.

Contemporary *Leptomyrmex* ants and *Mastotermes* termites coexist in tropical forests in northern Australia and surrounding areas. These forests are classified as moist to dry tropical forests, very similar to the biome of the departed forest,[55] further indicating that stenotopic animals seek out their mandatory habitat.

This book has given a fleeting glimpse—literally only a few seconds' worth—of an extinct tropical moist forest based on information known from organisms in Dominican amber. Ironically, even as this is being read, such tropical moist forests are being destroyed at an alarming rate worldwide. The biota of these modern-day forests face the same destiny as those in the amber forest: extinction. It has been said that there may not be adequate time to even begin cataloging or preserving all the life forms in these modern forests before they have vanished.

Will the original flora and fauna of today's tropical forests ultimately have to be contemplated through fossils in copal and amber some thousands or millions of years from now, by hopefully a wiser and more enlightened human race, so that mankind can learn about the life they destroyed?

General Conclusions

We have shown how it is possible to reconstruct, from plant and animal remains in amber, the basic type of forest that existed millions of years ago in the Dominican Republic. It is also possible to determine the ecology and even to reconstruct interactions that occurred between organisms in this forest by using characteristics of the modern-day descendants of the fossils. Comparing the biota of the past forest with that occurring in Hispaniola today demonstrates how important past climatic patterns are in determining the present distribution of plants and animals. This study also demonstrates the long-term stability of host-parasite, predatory-prey, and symbiotic associations and in so doing, supports the principle of behavioral fixity in the fossil record.

4

Amber Today

Handling, Photographing, and Preserving Amber

Amber purchased from dealers has normally already been polished by craftsmen in the Dominican Republic. Thus, any inclusion can easily be detected by examining the sample with a hand lens or under the low power of a binocular microscope (from 10 to 30X). However, the specimens may not be orientated correctly for scientific study or photography, and the pieces then need to be reshaped. If more than one specimen is in the nugget, the piece should first be cut to separate them, as long as they are not too close together.

For small and medium pieces, cutting can be done with slow and deliberate movements using a firmly held fine jeweler's hand saw. Larger pieces can be cut using a high speed trim saw equipped with a diamond blade at speeds up to 4,200 rpm. Remember that older amber as well as some Dominican material may contain fractures, and shear forces generated by a saw may shatter the piece.

Once the specimen has been isolated, the piece is ready to be reshaped using various grades of sandpaper. The surfaces to be modified depend on which portions of the specimen need to be seen (dorsal, ventral, or lateral). Rough surfaces from the saw blade can be removed with a 200 grit paper, followed with 400 grit. Again, sanding can be done by hand or on any of the commercially available belt sander units equipped with a water cooling system. Water not only removes the dust, but also keeps the piece from overheating and fracturing or glazing.

After sanding, the piece is ready to be polished. Again, this can be accomplished by hand, using a piece of flannel, with toothpaste as an abrasive. However, for larger pieces, a standard buffing or polishing wheel (3–6 inches in diameter) is much quicker. Any abrasive that is recommended for plastic will work for amber, although different hardnesses should be used under different temperature regimes. With motorized units, a final hand polishing may be necessary to remove deposits of polishing compound on the amber surface. Remember to wear protective glasses when working with any motorized unit. A buffing wheel can catch on a bur or the edge of a piece and whirl it toward you at considerable speeds. If this does happen, the piece is often broken as it strikes the wall or floor. Various kinds of glue, especially the fast-drying ones, can be used to reattach the portions. When dry, these glues are so strong that additional sanding and polishing, which is often necessary, can be done even with electric units without further separation along the break line.

For photographing specimens, it is often helpful to reshape the piece as described above. A flat surface is desirable for photography, but if that is impossible, it may be feasible to place the specimen in a liquid, which lowers the appearance of surface aberrations. Best results can be obtained by immersion in various oils, but care should be taken here. The liquid will enter any small cracks that are present, and, if it reaches the specimen, it will immediately darken and the surface and internal tissues will be modified irrevocably.

The preservation of amber is a topic of much concern to private collectors as well as collection managers and curators in public institutions. Most important is to insure that a collection be shielded from fluctuating temperatures, light, and air as much as possible. A constant temperature is important since amber is a poor conductor of heat. A move from freezing temperatures into a warm room will often damage large pieces of amber: they can crack right in front of your eyes as the heat enters the matrix. Most light, especially sunlight, will heat the specimen, which is why amber should be displayed under cold light. Exposure to air (oxygen) over time will darken the surface, often imparting a reddish color. Eventually, crazing and cracking will occur, especially with periodic temperature fluctuations.

The effect of humidity on amber still has to be investigated. Our personal observations indicate that a humid environment appears to be better than a dry one. Many have attempted to preserve specimens by storing them in solutions of water, glycerin, or corn syrup, or embedding the amber in compounds like Canada balsam or in various plastics. The latter may be the best way of protecting amber over the years but it may hinder viewing the specimen. Pieces that we and others embedded in bioplastic over the past twenty years have not altered in appearance, so the tedious process may have its merits though it certainly is not practical for art objects. Coating the latter with liquid waxes and compounds that harden and provide a thin layer of protection for normal viewing is a practical solution.

Amber Imitations and Forgeries

Over the ages, a number of materials, such as glass, plastics, horn, bone, stone, jet, and copal, have been used to imitate amber. There is even an old Chinese recipe that shows how to make "amber" out of eggs, vinegar, and oil. Of these imitations, plastic and copal concern collectors the most since they often contain large, rare specimens that could be scientifically valuable if they were in real amber.

However, amber forgeries do not always involve organisms that are intentionally placed in a substitute. Shops in America and abroad are filled with copal from Colombia and bordering countries in South America, most of which is under five hundred years of age[1] and is labeled as either Colombian, Dominican, Mexican, or even Baltic amber. A quick hot point or solvent test will normally expose the soft nature of the material. Colombian copal is highly fossiliferous with naturally occurring invertebrates and plants, and can be quite attractive. Copal from Madagascar and East Africa is also being sold as amber by some dealers.

Some forgers have learned how to embed larger organisms in the matrix, which introduces the second category of imitations, intentional fakes. Such pieces may have started just as an amusement or to fool friends, but many turned into products of a profitable hobby that could have unfortunate consequences for scientists. Intentional fakes are usually made with copal and plastic. Copal is a natural resin that is semifossilized and ranges in age anywhere from ten (as in some Colombian copal) to forty thousand years (as with some New Zealand copal, better known as kauri gum). A general definition for copal is a resin that is no longer sticky, cannot be molded by hand, and usually is hard enough to be cracked. Copal can be collected directly from the bark of a tree but is normally in the ground under the tree source, sometimes fairly deep in the area of a former forest. Because most copal is under forty thousand years old, it can be radiocarbon dated. Several laboratories in the United States will date such samples for the general public (at a cost).

Most copal used in fakes comes from Africa, Madagascar, and South America and is produced by leguminous trees. This material is generally lighter in color than most amber. Copal from New Zealand or Australasia, however, produced by kauri trees, exhibits a wide range of hues, from yellow to nearly black.

Because the molecules are not as tightly bound as in amber, the surface of copal can be modified with organic solvents like ether, acetone, or alcohol. With the hot point method (a red-hot needle placed against the outer surface), the surface can be melted easily, emitting white, resinous fumes. Because of its young age, copal can be softened with heat and various organisms placed inside.

Another choice selection for amber imitations is plastic. The forger has a wide variety of plastics from which to choose and can color the matrix to imitate real amber.[2] Bubbles, debris, and other material can be added to make the piece appear authentic. Any piece of "amber" containing a vertebrate or a large, showy invertebrate (such as centipedes, millipedes, cockroaches, social wasps, and butterflies) should be regarded as suspect until it passes a series of tests. The hot point test on plastics, usually produces a synthetic odor that is recognized instantly by the trained nose. The solvent test can expose some of the softer plastics but not the harder varieties. The salt solution or specific gravity test (dissolving two and a half level tablespoons of table salt in a cup of water) will catch the heavier plastics, which sink, but the lighter plastics often float like real amber. Of course laboratories are equipped with modern methods (nuclear magnetic resonance, infra-red spectroscopy, pyrolysis-mass spectroscopy, etc.) that can immediately recognize the plastic origin of a material by analyzing a small sample.

Perhaps the most difficult forgeries to detect, which have fooled even the specialists, are those where a fossil is inserted into a cavity made inside a piece of authentic amber. Usually a piece of amber is broken with a clean fracture, a small depression is made in the center of one of the broken portions, an organism is inserted into the depression, the cavity is filled with a plastic or copal of the same color, and the pieces are glued back together again. None of the above described testing methods can detect such a forgery. Only a sharp eye that can percieve the slight imperfections made when the cavity was filled, or a specialist recognizing that the specimen is of modern age (true amber fossils almost always contain extinct species), can catch these fakes.

Tissue Preservation and Ancient DNA in Amber

Ancient DNA represents nucleic acids obtained from dead organisms that have been preserved for at least twenty-five years. Research in this field was propelled into the limelight with the report by Berkeley scientists in 1984 that DNA had been recovered from a 140-year-old museum specimen of the extinct quagga, a zebra-like animal native to South Africa.[3] Enough sequences were obtained to show that the quagga was more closely related to the zebras than the horses. Thus not only was ancient DNA recovered, but it was used to demonstrate the phylogenetic relationship of an extinct animal. With the invention of the polymerase chain reaction, where small traces of DNA characteristic of ancient material can be amplified to workable quantities, scientific work in this area became much more feasible and the discipline grew rapidly.

Back in 1982, we sectioned a piece of amber containing a small fungus gnat and discovered for the first time that the intracellular components of body cells had been preserved for 40 million years.[4] Still present in a recognizable state were epidermal cell nuclei, lipid droplets, mitochondria, ribosomes, and other cellular components, revealing a degree of preservation never before observed in fossilized tissues. Publication of this discovery fermented news articles speculating on the possibility of resurrecting ancient creatures from DNA located within these cells. While bringing extinct animals back to life seems highly unlikely, ancient DNA recovered from entombed biota could be used to examine the diversity of life that flourished in the ancient amber forest. In 1992, ancient DNA was extracted from the first amber fossil, a stingless bee in Dominican amber.[5] This was followed by a termite,[6] an algarrobo leaf in Dominican amber,[7] and a weevil in Lebanese amber.[8] The results of these studies were used to answer questions regarding biogeography, evolution, and phylogenetic relationships. Nucleic acids recovered from the ancient algarrobo leaf showed that the closest living relative was a relict species occurring along the eastern coast of Africa. Later, ultrastructural studies on a similar algarrobo leaf revealed that chloroplasts within the cells had their internal components still intact.[9] Amber is full of bits and pieces of broken leaves, bark, wood, flowers, and seeds, all of which are too fragmentary to be identified by standard means. However, they could possibly be identified from the fragments of DNA that reside in their cells. First, however, scientists need to accumulate a molecular database on the vast majority of extant tropical plants so that matches can be made. Aside from aiding in the identification of ancient flora and fauna, the analysis of DNA from fossil inclusions is an excellent way of studying evolution at the molecular level.

However, successful attempts to remove fossil DNA are not universal, and recent failures demonstrate that a set of criteria, listed below, needs to be fulfilled if success is to be expected:

1. Tissue extraction should be carried out in a laboratory designed for molecular studies, especially one in which successful results in obtaining fossil DNA have already been achieved.
2. All procedures should be conducted by personnel experienced with handling ancient DNA or at least have already successfully extracted DNA from modern tissue.
3. Amber specimens subjected to DNA analysis should be in pristine morphological condition, showing no signs of decay or contamination and should not have previously been exposed to high temperatures or other conditions that could degrade any remaining DNA (many Baltic amber specimens are now autoclaved to improve the visibility of the inclusion; such pieces would not be suitable for ancient DNA studies).
4. Only extraction procedures which do not subject the tissue to further physical or chemical damage should be used.
5. Contamination should be less than 10 percent—if greater, it is an indication that there is a sterility or contamination problem somewhere within the system.

Negative results and contamination do not necessarily mean that DNA is absent, only that there may be a technical problem somewhere. A few researchers have refused to accept earlier reports of nucleic acid recovery from amber because their own experiments on DNA instability demonstrated that decay would be complete after 100,000 years and it would be no use to look in older tissue. While such figures may be valid for excised tissue or cells held under laboratory conditions or directly exposed to the environment, it does not apply to specialized situations like amber fossils, where the tissue is protected for millions of years by continuous contact with resin. Several studies have shown that amber provides conditions conducive to the long-term preservation of amino acids.[10] Since the breakdown of amino acids (racemization) and the decay of nucleic acids (depurination) appear to be affected by the same conditions over time (moisture, temperature, and the attachment of metal ions to proteins), the preservation of one is an indication of the preservation of the other.

Previous investigations have shown that DNA from amber organisms is fragmented, degraded, and strongly cross-linked with other molecules in the cell. Extraction methods that take this into consideration will be more successful than those that don't. In the future, it may be possible to repair some of the ancient DNA, which would allow the retrieval of longer, more authentic sequences.

While ultrastructural studies demonstrate the presence of cells and cellular inclusions in locations similar to those in present-day tissues, it should be emphasized that many amber arthropods may have remained on the surface of the resin for hours and undergone internal decay, which occurs quite rapidly under tropical conditions. Only those forms that were completely immersed after contacting the resin are likely to retain extractable DNA. It may be difficult to distinguish between invertebrates that have decomposed internally from those with well-preserved tissues after 20–40 million years.

Appendixes

Appendix A. Frequency of Organisms in Dominican Amber

From an examination of 3,017 random pieces with inclusions
(obvious rare pieces had already been removed)

Arthropoda (arthropods)
  Arachnida (arachnids)
    Acari (mites)   3
    Araneae (spiders)   88
  Crustacea
    Isopoda (isopods)   2
  Myriapods
    Diplopoda (millipedes)   5
    Chilopoda (centipedes)   1
  Hexapoda (insects and related forms)
    Collembola (springtails)   21
    Archeognatha (bristletails)   8
    Blatteria (cockroaches)   3
    Orthoptera (grasshoppers)   1
    Grylloptera (crickets)   17
    Isoptera (termites)   86
    Psocoptera (bark lice)   173
    Thysanoptera (thrips)   6
    Homoptera
      Fulgoroidea (planthoppers)   51
      Cicadellidae (leafhoppers)   43
    Hemiptera (true bugs)   13
    Coleoptera (beetles)
      Platypodidae (flat-footed beetles)   115
      Other beetles   86
    Trichoptera (caddisflies)   8
    Lepidoptera
      Moths   60
    Diptera (flies)
      Acalyptratae (acalyptrate muscoid flies)   149
      Anisopodidae (wood gnats)   11
      Cecidomyiidae (gall midges)   197
      Ceratopogonidae (biting midges)   81
      Chironomidae (midges)   89

Mycetophilidae (fungus gnats)   110
Phoridae (phorid flies)   34
Psychodidae
    (moth flies)   83
    (sand flies)   14
Scatopsidae (scavenger flies)   124
Sciaridae (dark-winged fungus gnats)   57
Tipulidae (crane flies)   6
other unidentified flies   62
Hymenoptera (ants, wasps, etc.)
    Apidae (bees)
        (stingless bees)   156
    Formicidae (ants)
        (worker ants)   497
        (winged adult ants)   286
    Parasitica (parasitic wasps)   168
Unidentified insect larvae   10
Plants
    Portions of moss   3
    Portions of liverworts   6
    Angiosperm remains
        Bark pieces   1
        Buds   6
        Bud scales   2
        Flowers   5
        Leaves   8
        Petals   3
        Developing fruit (pod)   1
        Sepals   13
        Stamens   13
        Stems   5
        Unidentified   32

Appendix B. Biota in Dominican Amber

(Citing genera [extinct genera are in **bold print**], families, and higher
categories. This list contains the majority of biota identified thus far)

Fungi
  Basidiomycetes
    Coprinaceae
      Coprinites
  Ascomycetes
    Lecanorales
      Parmelites
    Xylariadeae
      Xylaria
    Hyphomycetes
      Aspergillis
      Beauveria
      Geotrichites
    Entomophthorales
Pteridophyta
Ficales
  Grammaticeae
    Grammatis
Bryophytes
  Hepaticae
    Frullaniaceae
      Frullania
    Lejeuneaceae
      Archilejeunea
      Blepharolejeunea
      Bryopteris
      Ceratolejeunea
      Cyclolejeunea
      Cyrtolejeunea
      Drepanolejeunea
      Lejeunea
      Leucolejeunea
      Lopholejeunea
      Marchesinia
      Mastigolejeunea
      Neurolejeunea
      Prionolejeunea

      Stictolejeunea
    Lepidoziaceae
      Bazzania
    Radulaceae
      Radula
  Musci
    Adelothediaceae
      Adelothecium
    Calymperaceae
      Calymperes
      Syrrhopodon
    Hypnaceae
      Hypnum
      Mittenothamnium
    Lembophyllaceae
      Pilotrichella
    Leucobryaceae
      Octoblepharum
    Neckeraceae
      Homalia
      Neckera
    Sematophyllaceae
      Clastobryum
    Thuidiaceae
      Thuidium
Angiospermae
  Amaranthaceae
  Ancistrocladaceae
  Apocynaceae
  Bignoniaceae
  Bromeliaceae
  Euphorbiaceae
  Gramineae
    Arthrostylidium
    Panicum
    Pharus
  Hippocrateaceae

Peritassa
Lauraceae
Nectandra
Leguminoseae
Acacia
Hymenaea
Mimosa
Peltogyne
Prioria
Lythraceae
Meliaceae
Swietenia
Trichilia
Moraceae
Ficus
Myristiaceae
Virola
Orchidaceae
Palmaceae
Roystonea
Sapindaceae
Solanaceae
Thymeliaceae
Ulmaceae
Trema
Urticaceae
Protozoa
Gromiida
Cyphoderiidae
Cyphoderia
Physarales
Bacteria
Actinomycetes
Nematomorpha
Nematoda
Tylenchida
Allantonematidae
Parasitylenchus
Iotonchiidae
Iotonchium
Mermithidae
Heydenius

Rhabditida
Rhabditidae
Rotifera
Bdelloida
Habrotrochidae
Habrotrocha
Mollusca
Gastropoda
Ferussaciidae
Helicinidae
Oleacinidae
Varicella
Spiraxidae
Spiraxis
Strobilopsidae
Strobilops
Subulinidae
Subulina
Prosobranchia
Annelida
Enchytraeidae
Tardigrada
Crustacea
Amphipoda
Talitridae
Tethorchestia
Isopoda
Philoscidae
Platyarthidae
Trichorhina
Pseudarmadillidae
Pseudarmadillo
Sphaeroniscidae
Protosphaeroniscus
Styloniscidae
Diplopoda
Chelodesmidae
Chytodesmidae
Dasyodesmidae
Dasyodesmus
Glomeridesmidae
Glomeridesmus

Inodesmidae
Inodesmus
Lophoproctidae
Lophoproctus
Polydesmidae
Polyxenidae
Pseudononnolenidae
Epinannolene
Pyrgodesmidae
Docodesmus
Iomus
Siphonophoridae
Lophodesmus
Psochodesmus
Siphonocybe
Siphonophora
Siphonotidae
Siphonotus
Stemmiulidae
Prostemmuilus
Chilopoda
Geophilidae
Scolopendrida
Cryptopidae
Cryptops
Onychophora
Symphyla
Scutigerellidae
Scutigerella
Hexapoda
Diplura
Anajapygidae
Procampodiidae
Collembola
Entomobryidae
Cyphoderus
Lepidocyrtus
Paronella
Pseudosinella
Salina
Seira
Isotomidae

Cryptopygus
Isotoma
Sminthuridae
Sphyrotheca
Aracheognatha
Machilidae
Meinertellidae
Neomachilellus
Thysanura (Zygentoma)
Ateluridae
Archeatelura
Lepismatidae
Ctenolepisma
Nicoletiidae
Trinemurodes
Ephemeroptera
Baetidae
Cloedes
Emphemeridae
Leptophlebiidae
Borinquena
Careospina
Odonata
Coenagrionidae
Diceratobasis
Plecoptera
Perlidae
Dominiperla
Blatteria
Blattellidae
Euthyrrhaphidae
Grylloptera
Mogoplistidae
Ornebius
Trigonidiidae
Abanaxipha
Anaxipha
Grossoxipha
Proanaxipha
Mantodea
Mantidae
Phasmida

Phasmatidae
Orthoptera
  Acrididae
  Eumastacidae
  Tettigonidae
  Tridactylidae
Dermaptera
Isoptera
  Kalotermitidae
    Cryptotermes
  Mastotermitidae
    Mastotermes
  Rhinotermitidae
    Coptotermes
  Termitidae
    Nasutitermes
Embioptera
  Anisembiidae
    Mesembia
  Teratembiidae
    Oligembia
Zoraptera
  Zorotypidae
    Zorotypus
Psocoptera
  Amphientomidae
    Lithoseopsis
  Archipsocidae
    Archipsocus
  Caeciliidae
    Enderleenella
    Xanthocaecilius
  Cladiopsocidae
    Cladiopsocus
    Spurostigma
  Dolabellopsocidae
    Isthmopsocus
  Epipsocidae
    Epipsocus
  Hemipsocidae
    Hemipsocus
  Lepidopsocidae

Echmepteryx
Liposcelidae
  Liposcelis
Philotarsidae
  Broadheadia
Polypsocidae
Pseudocaeciliidae
Psocidae
  Blaste
  Blastopsocus
  Indiopsocus
  Psyllipsocus
  Ptycta
  Trichadenotecnum
Psoquillidae
  Rhyopsocus
Ptiloneuridae
Thysanoptera
Homoptera
  Aetalionidae
  Aleyrodidae
  Cercopidae
  Cicadellidae
    Eualebra
    Krisna
    Xestocephalus
  Cicadidae
  Cixiidae
  Coccidae
  Delphacidae
  Derbidae
  Diaspididae
  Dictyopharidae
  Dipsocoridae
  Drepanosiphidae
    Mindazerius
  Eurymelidae
  Flatidae
  Fulgoridae
    Krocorites
  Issidae
  Kinnaridae

Membracidae
Ortheziidae
Pseudococcidae
Psyllidae
Hemiptera
  Alydidae
  Anthocoridae
  Aradidae
    Calisiopis
    Eretmocoris
  Berytidae
  Cydnidae
  Dipsocoridae
  Enicocephalidae
    Alienates
    Enicocephalus
    Paralienates
  Gerridae
    Electrobates
  Isometopidae
  Leptopodidae
  Lygaeidae
  Miridae
    Diphleps
    Reuteroscopus
    Stomatomiris
  Nabidae
  Pentatomidae
  Reduviidae
    Alumeda
    Apicrenus
    Apiomerus
    Empicoris
    Empiploiariola
    Malacopus
    Paleoploiariola
    Praecoris
    Rasahus
  Rhopalidae
  Saldidae
    Leptosalda
  Schizopteridae

Termitaphididae
  Termitaradus
Thaumastocoridae
  Paleodoris
Tingidae
Veliidae
  Halovelvia
Neuroptera
  Ascalaphidae
  Chrysopidae
    Leucochrysa
  Coniopterygidae
    Coniopteryx
    Spiloconis
  Hemerobiidae
    Hemerobius
    Notiobiella
  Mantispidae
    Mantispa
  Myrmeliontidae
    Porrerus
  Sialidae
Coleoptera
  Alleculidae
    Allecula
    Hymenorus
  Anobiidae
  Anthicidae
  Biphyllidae
  Bostrichidae
  Brachypsectridae
  Brentidae
  Bruchidae
    Caryobruchus
  Buprestidae
  Byturidae
  Cantharidae
  Carabidae
    Aephnidius
    Arthrostictus
    Bembidion
    Eohomopterus

Homopterus
Pachyteles
Phloeoxena
Protopaussus
Selenophorus
Stenognathus
Cerambycidae
Eburia
Elaphidion
Merostenus
Methia
Pentanodes
Pentomacrus
Plectromerus
Ceratocanthidae
Cerophytidae
Cerylonidae
Chrysomelidae
Chalcosicya
Colaspoides
Cryptocephalus
Diachus
Glyptoscelis
Leptonesiotes
Metachroma
Sceloenopla
Walterianella
Ciidae
Cleridae
Coccinellidae
Colydiidae
Colydium
Corylophidae
Cryptophagidae
Cucujidae
Brontes
Telephanus
Curculionidae
Apion
Anthonomus
Catolethrus
Dryophthorus

Himatium
Rhinanisus
Tyloderma
Dascillidae
Dermestidae
Discolomidae
Dryopidae
Dytiscidae
Copelatus
Elateridae
Endomychidae
Erotylidae
Eucinetidae
Eucnemidae
Euglenidae
Haliplidae
Helodidae
Histeridae
Hydrophilidae
Anacaena
Inopeplidae
Lagriidae
Lampyridae
Languriidae
Lathridiidae
Leiodidae
Limnichidae
Limulodidae
Lucanidae
Lycidae
Lyctidae
Lymexylidae
Atractocerus
Melandryidae
Melyridae
Micromalthidae
Micromalthus
Monommatidae
Mordellidae
Mycetophagidae
Litargus
Mycteridae

Nitidulidae
Noteridae
Ostomidae
Pedilidae
Phalacridae
Platypodidae
   Cenocephalus
   Tessocerus
Pselaphidae
   Apharus
   Arthmius
   Barroeuplectoides
   Brachyglutina
   Jubus
   Scalenarthrus
   Thesium
   Trimiena
Ptiliidae
Ptilodactylidae
Ptinidae
Pyrochroidae
Rhipiphoridae
Rhizophagidae
Rhysodidae
Salpingidae
Scaphidiidae
Scarabaeidae
   Canthidium
   Rhyparus
   Termitodius
Scolytidae
   Cladoctonus
   Cnemonyx
   Corthylites
   Dryomites
   Gnathotrichus
   Hypothenemus
   Microcites
   Microcorthylus
   Paleophthorus
   Paleosinus
   Phloeotribus

Pityophthorus
Protosinus
Pycnarthrum
Scolytus
Scraptia
Scraptiidae
Scydmaenidae
   Euconnus
   Homoconus
   Mastigonius
   Microscydmus
   Neuraphes
Silvanidae
   Platamus
Staphylinidae
   Neoxantholinus
   Prorhinopsenius
   Tachyporus
Tenebrionidae
   Cymatothes
   Hesiodobates
   Hypogena
   Lioderma
   Lobopoda
   Lorelus
   Neomida
   Nesocyrtosoma
   Parahymenorus
   Platydema
   Rhipidandrus
   Stitira
   Trientoma
   Tyrtaeus
   Wattius
Throscidae
Trogidae
Strepsiptera
Bohartillidae
   Bohartilla
Elenchidae
   Protelencholax
Myrmecolacidae

Caenocholax
Myrmecolax
Stichotrema
Trichoptera
Atopsylidae
Atopsyche
Glossosomatidae
Campsiophora
Cubanoptila
Helicopsychidae
Helicopsyche
Hydropsychidae
Leptonema
Palaehydropsyche
Hydroptilidae
Alisotrichia
Leucotrichia
Ochrotrichia
Oxyethira
Leptoceridae
Nectopsyche
Setodes
Philopotamidae
Chimarra
Polycentropodidae
Antillopsyche
Cernotina
Lepidoptera
Acrolophidae
Blastobasidae
Cosmopterygidae
?Anoncia
?Pyroderces
Epipyropidae
Gelechiidae
Geometridae
Gracillariidae
Noctuidae
Nymphalidae
Oecophoridae
Psychidae
Pyralidae

Riodinidae
Napea
Theope
Tineidae
Tortricidae
Polyvenia
Diptera
Acroceridae
Ogcodes
Agromyzidae
Anisopodidae
Mesochira
Mycetobia
Valeseguya
Anthomyidae
Anthomyza
Coenosopsites
Asilidae
Leptogaster
Ommatius
Wilcoxia
Asteiidae
Asteia
Loewimyia
Aulacigastridae
Aulacigaster
Bibionidae
Delophus
Plecia
Bombylidae
Glabellula
Mythecomyia
Poecilognathus
Calliphoridae
Carnidae
Moneura
Cecidomyiidae
Ceratopogonidae
Baeodasymyia
Culicoides
Heteromyia
Palpomyia

Phaenobezzia
Chamaemyiidae
Chaoboridae
  Dayomyia
Chironomidae
Chloropidae
  Tricimba
Clusiidae
  Clusiodes
Conopidae
  Stylogaster
Culicidae
  Aedes
  Anopheles
  Culex
  Orthopodomyia
Cypselosomatidae
  Latheticomyia
Dixidae
Dolichopodidae
  Condylostylus
  Diostracus
Drosophilidae
  Chymomyza
  Drosophila
  Hyalistata
  Miomyia
  Neotanygastrella
  Protochymomyza
  Scaptomyza
  Stegana
Empididae
  Rhamphomyia
Ephydridae
  Beckeriella
Lauxaniidae
Lestremiidae
Longichetidae
Micropezidae
  Raineria
Milichiidae
  Meonura

Pholeomyia
Phyllomyza
Muscidae
Mycetophilidae
  Allodia
  Aphrastomyia
  Boletina
  Celosia
  Cordyla
  Euceratoplatus
  Exechia
  Exechiopsis
  Keraplatus
  Leia
  Macrocera
  Megophthalmedia
  Metaleia
  Mycetophila
  Mycomyia
  Orfelia
  Phronia
  Platyura
  Proceroplatus
  Rhimozia
  Sciaphita
  Synaphe
  Tetragoneura
  Trichonta
Odiniidae
  Odinia
Otitidae
Pachyneuridae
Periscelidae
Phoridae
  Abaristophora
  Dohrniphora
  Megaselia
  Metopina
  Puliciphora
Pipunculidae
Platypezidae
Psychodidae

Lutzomyia
Nemopalpus
Psychoda
Telmatoscopus
Trichomyia
Rachioceridae
Rhagionidae
Chrysopilus
Richardiidae
Scatopsidae
Rhegmoclemina
Sciaridae
Scenopinidae
Metatrichia
Sciomyzidae
Simuliidae
Sphaeroceridae
Stratiomyidae
Nothomyia
Synneuridae
Syrphidae
Copestylum
Tabanidae
Stenotabanus
Tachinidae
Tanypezidae
Tephritidae
Ceratodacus
Protortalotrypeta
Therividae
Tipulidae
Austrolimnophila
Dicranomyia
Elephantomyia
Erioptera
Geranomyia
Gnophomyia
Gonomyia
Limonia
Molophilus
Polymera
Rhipidia

Styringomyia
Toxorhina
Trentepophila
Trichoceridae
Trixoscelididae
Siphonaptera
Pulicidae
Pulex
Rhopalopsyllidae
Rhopalopsyllus
Hymenoptera
Andrenidae
Heterosaurus
Aphelinidae
Aphelinus
Coccophagus
Apidae
Proplebeia
Paleoeuglossa
Argidae
Didymia
Bethylidae
Parasierola
Braconidae
Aivalykus
Apanteles
Bracon
Choeras
Cyclostoma
Diolcogaster
Macrocentrus
Phanerotoma
Ceraphronidae
Ceraphron
Chalcididae
Brachymeria
Psilochalcis
Spilochalcis
Colletidae
Chilicola
Chrysididae
Cynipidae

Diapriidae
  Basalys
  Corynopria
  Trichopria
Dryinidae
  Alphadryinus
  Aphelopus
  Bocchus
  Dryinus
  Thaumatodryinus
Encrytidae
  Apsilophrys
  Copidosoma
Eulophidae
  Euderus
Eupelmidae
  Zaischnopsis
Evaniidae
Figitidae
Formicidae
  Acanthognathus
  Acanthostichus
  Anochetus
  Aphaenogaster
  Apterostigma
  Azteca
  Brachymyrmex
  Camponotus
  Cephalotes
  Crematogaster
  Cylindromyrmex
  Cyphomyrmex
  Dendromyrmex
  Diplorhoptrum
  Discothyrea
  Dolichoderus
  Erebomyrma
  Eurhopaleothrix
  Gnamptogenys
  Hypoclinea
  Hypoponera
  Ilemomyrmex

  Leptanilloides
  Leptomyrmex
  Leptothorax
  Linepithema
  Monacis
  Myrmelachista
  Neivamyrmex
  Octostruma
  Oligomyrmex
  Oxyidris
  Pachycondyla
  Paraponera
  Paratrechina
  Pheidole
  Platythyrea
  Pogonomyrmex
  Prenolepis
  Prionopelta
  Proceratium
  Pseudomyrmex
  Smithistruma
  Solenopsis
  Strumigenys
  Tapinoma
  Trachymesopus
  Trachymyrmex
  Wasmania
Halictidae
  Eickwortapis
  Oligochlora
  Neocorynura
Ichneumonidae
Megaspilidae
  Conostigmus
  Dendrocerus
Mutillidae
  Dasymutilla
Mymaridae
  Camptoptera
  Palaeopatasson
Orussidae
Platygasteridae

Acerotellas
Pompilidae
Proctotrupidae
Pteromalidae
   Spalangiopelta
Scelionidae
   Baeus
   Boryconis
   Calliscelio
   Calotelea
   Ceratobaeus
   Gryon
   Holoteleia
   Idris
   Oethecoctonus
   Probaryconus
   Scelia
Sclerogibbidae
Scolebythidae
   Clystopsenella
   Dominibythus
Sphecidae
   Crabro
   Crossocerus
   Nitella
   Nyssonia
   Pison
   Trypoxylon
Tiphiidae
Torymidae
   Neopalachia
   Zophodetus
Vespidae
   Agelaia
Arachnida
Acari
  Argasidae
   Ornithodoros
  Bdellidae
  Caeculidae
   Procaeculus

Carabodidae
  Carbodes
  Phyllocarabodes
Ceratozetoidea
Cunaxidae
Eremaeozetidae
  Eremaeozetes
Erythraeidae
  Fallopia
Galumnidae
  Galumna
Gamasidae
Hermanniellidae
  Sacculobates
Ixodidae
  Ambylomma
Liacaridae
Liodidae
  Liodes
  Teleioliodes
Listrophoridae
Macrochelidae
Mochlozetidae
  Mochlozetis
Oppidae
  Oppia
Oribatidae
Oribotritiidae
  Oribotritia
Oripodidae
  Oripoda
Otocepheidae
  Dolicheremaeus
Plateremaeoidea
Scutoverticidae
  Arthovertex
Smarididae
  Hirstiostoma
  Smaris
Trhypochthoniidae
  Allonothrus

Trombididae
Araneae
  Agelenidae
  Amaurobiidae
  Anapidae
    Palaeoanapis
  Anyphaenidae
    Aysha
    Teudis
    Wulfila
  Araneidae
    Araneometa
    Araneus
    Cyclosa
    Fossilaraneus
    Pycnosinga
  Barychelidae
    Psalistops
  Caponiidae
    Nops
  Clubionidae
    Clubionoides
    Strotarchus
  Ctenidae
    Nanoctenus
  Ctenizidae
    Bolostromus
  Dictynidae
    Hispaniolyna
    Palaeodictyna
    Palaeolathys
    Succinyna
  Dipluridae
    Ischnothele
    Masteria
    Microsteria
  Gnaphosidae
    Drassyllinus
  Hahniidae
  Hersiliidae
    Tama

Heteropodidae
  Tentabuna
Linyphiidae
  Agyneta
  Lepthyphantes
  Palaeolinyphia
Lioncranidae
Microstigmatidae
Mimetidae
  Mimetus
Myrmeciidae
  Castianeira
  Chemmisomma
  Corinna
  Megalostrata
  Veterator
Nesticidae
  Hispanonesticus
Ochyroceratidae
  Arachnolithulus
Oecobiidae
  Oecobius
Oonopidae
  Fossilopaea
  Gamasomorpha
  Heteroonops
  Opopaea
  Orchestina
Oxyopidae
  Oxyopes
Palpimanidae
  Otiothops
Philodromidae
Pholcidae
  Modisimus
  Pholcophora
  Serratochorus
Pisauridae
Pychothelidae
Salticidae
  Corythalia

Descangeles
Descanso
Lyssomanes
Nebridia
Pensacolatus
Phlegrata
Thiodina
Scytodidae
  Scytodes
Segestriidae
  Ariadna
Selenopidae
  Selenops
Sicariidae
  Loxosceles
Symphytognathidae
Tetrablemmidae
  Monoblemma
Tetragnathidae
  Azilia
  Cyrtognatha
  Homalometa
  Leucauge
  Nephila
  Tetragnatha
Theraphosidae
  Ischnocolinopsis
Theridiidae
  Achaearanea
  Anelosimus
  Argyrodes
  Chrosiothes
  Chrysso
  Corjutidion
  Craspedisia
  Dipoenata
  Episinus
  Lasaeola
  Spintharus
  Stemmops
  Styposis
  Theridion

Theridiosomatidae
  Palaeoepeirotypus
  Theridiosoma
Thomisidae
  Heterotmarus
  Komisumena
Uloboridae
  Miagrammopes
Opiliones
  Minuidae
    Kimula
  Phalangodidae
    Pellobunus
    Philacarus
  Samoidae
    Hummelinckiolus
Pseudoscorpiones
  Cheiridiidae
    Cryptocheiridium
  Cheliferidae
    Parawithius
  Chernetidae
    Americhernes
    Pachychernes
  Chthoniidae
    Lechytia
    Pseudochthonius
Scorpiones
  Buthidae
    Centruroides
    Microtityus
    Tityus
Schizomida
Solpugida
  Ammotrechidae
    Happlodontus
Ambypygi
  Phrynidae
    Phrynus
Amphibia
  Anura
    Leptodactylidae

Eleutherodactylus
Reptilia
  Squamata
    Gekkonidae
      Sphaerodactylus
    Iguanidae
      Anolis
  Serpentes
    Typhlopidae
Aves
  Piciformes

Picidae
  Nesoctites
Mammalia
  Rodentia
    Capromyidae
      ? Plagiodontia
  Carnivora
  Chiroptera
  Insectivora
    Nesophontidae
      Nesophonites

Chapter References

Chapter 1. Introduction

1. Lewis, J. F., Draper, G., Bowin, C., Bourdon, L., Maurrasse, F., & Nagle, F. 1990. Hispaniola. In: The Caribbean Region, pp. 94–112. Eds. Dingo, G., & Case, J. E. Geology Society of America, Boulder, CO.
2. Ross, M. I., & Scotese, C. R. 1988. A hierachical tectonic model of the gulf of Mexico and Caribbean Region. Tectonophysics 155: 139–168.
3. Donnelly, T. W. 1992. Geological setting and tectonic history of Mesoamerica. In: Insects of Panama and Mesoamerica, pp. 1–13. Eds. Quintero, D., & Aiello, A. Oxford University Press, New York.
4. Poinar, Jr., G. O. 1992. Life in Amber. Stanford University Press, Stanford, CA.
5. Lambert, J. B., Frye, J. S., & Poinar, Jr., G. O. 1985. Amber from the Dominican Republic: Analysis by nuclear magnetic resonance spectroscopy. Archaeometry 27: 43–51.
6. Iturralde-Vincent, M. A., & MacPhee, R.D.E. 1966. Age and paleogeographic origin of Dominican amber. Science 273: 1850–1852.
7. Schlee, D. 1990. Das Bernstein-Kabinett. Stuttgarter Beitr. Naturkunde (C), no. 28.
8. Lewis et al., Hispaniola.
9. Boucot, A. 1990. Evolutionary Paleobiology of Behavior and Coevolution. Elsevier, Amsterdam.
10. Jacobs, M. 1981. The Tropical Rain Forest: A First Encounter. Springer-Verlag, Berlin.
11. Prance, G. T., & Elias, T. S., eds. 1977. Extinction is Forever. New York Botanical Garden, Bronx, NY.

Chapter 2. The Amber Forest

1. Poinar, Jr., G.O. 1991. *Hymenaea protera* sp. n. (Leguminosae, Caesalpinioideae) from Dominican amber has African affinities. Experientia 47: 1075–1082.
2. Lee, Y-T., & Langenheim, J. H. 1975. Systematics of the genus *Hymenaea* L. (Leguminoseae, Caesalpinioideae, Detarieae). University of California Press, Berkeley.
3. Stubblebine, W. H., & Langenheim, J. H. 1977. Effects of *Hymenaea courbaril* leaf resin on the generalist herbivore *Spodoptera exigua* (beet army worm). J. Chem. Ecol. 3: 633–647; Arrhenius, S. P., Foster, C. E., Edmonds,

C. G., & Langenheim, J. H. 1983. Inhibitory effects of *Hymenaea* and *Co-paifera* leaf resins on the leaf fungus *Pestalotia subcuticularis*. Biochem. Syst. Ecol. 11: 361–366.

4. Allen, P. H. 1956. The Rain Forests of Golfo Dulce. University of Florida Press, Gainesville.
5. Croat, T. B. 1978. Flora of Barro Colorado Island. Stanford University Press, Stanford, CA.
6. Allen, Rain Forests of Golfo Dulce.
7. del Valle, J. I. 1972. Introducion a la dendrologia de Colombia. Centro de Publicaciones de la Universidad Nacional de Colombia, Medellin; Allen, Rain Forests of Golfo Dulce.
8. Bolay, E. 1997. The Dominican Republic, a Country between Rain Forest and Desert: Contributions to the Ecology of a Caribbean Island. Margraf Verlag, Nordlingen, Germany.
9. Liogier, A. H. 1985. La Flora de la Espanola. 3. San Pedro de Macoris, Republica Dominicana.
10. Bolay, The Dominican Republic.
11. Jones, D. L. 1995. Palms throughout the World. Smithsonian Institution Press, Washington, DC.
12. Gentry, A. H. 1993. A Field Guide to the Families and Genera of Woody Plants of Northwest South America (Colombia, Ecuador, Peru) with Supplementary Notes on Herbaceous Taxa. University of Chicago Press, Chicago.
13. Ibid.
14. Hingston, R.W.G. 1932. A Naturalist in the Guiana Forest. Longmans, Green & Co., NY.
15. Poinar, Jr., G.O., & Columbus, J.T. 1992. Adhesive grass spikelet with mammalian hair in Dominican amber: First fossil evidence of epizoochory. Experientia 48:906–908.
16. Juddziesicz, E. J. 1987. Taxonomy and morphology of the tribe Phareae (Poaceae: Bambusoideae). Ph.D. diss., University of Wisconsin, Madison.
17. Gentry, Field Guide to Woody Plants.
18. Ibid.
19. Poinar, Jr., G .O., & Santiago-Blay, J. 1997. *Paleodoris lattini* gen. n., sp. n., a fossil palm bug, with habits discernable by comparative functional morphology. Entomol. Scand. 28: 307–310.
20. Gentry, Field Guide to Woody Plants.
21. Ramirez, W. 1970. Taxonomic and biological studies of neotropical fig wasps (Hymenoptera: Agaonidae). University of Kansas Science Bull. 49: 1–44.
22. Galil, J. 1985. Ficus. In: CRC Handbook of Flowering, vol. 6, pp. 331–349. Ed. Halevy, A. H. CRC Press, Boca Raton, Florida.

23. Poinar, Jr., G. O., & Herre, E. A. 1991. Speciation and adaptive radiation in the fig wasp nematode, *Parasitodiplogaster* (Diplogasteridae: Rhabditida) in Panama. Revue Nematol. 14: 361–374.

24. Poinar, Jr., G. O. 1996. A fossil stalk-winged damselfly, *Diceratobasis workii* spec. nov. from Dominican amber, with possible ovipositional behavior in tank bromeliads (Zygoptera: Coenagrionidae). Odonatologica 25: 381–385.

25. De Vries, P. J. 1997. The Butterflies of Costa Rica and Their Natural History, vol. 2: Riodinidae. Princeton University Press, Princeton, NJ.

26. Williams, N. H. 1982. The biology of orchids and euglossine bees. In: Orchid Biology, vol. 2, pp. 119–171. Ed. Arditti, J. Cornell University Press, Ithaca, NY.

27. Gradstein, S. R. 1993. New fossil hepaticae preserved in amber of the Dominican Republic. Nova Hedwigia 57: 353–374.

28. Poinar, Jr., G. O., and Ricci, C. 1992. Bdelloid rotifers in Dominican amber. Evidence for parthenogenetic continuity. Experientia 48: 408–410.

29. Poinar, Jr., G. O., & Singer, R. 1990. Upper Eocene gilled mushroom from the Dominican Republic. Science 248: 1099–1101.

30. Poinar and Ricci, Bdelloid rotifers.

31. Furth, D. G. 1988. The jumping apparatus of flea beetles (Alticinae)—the metafemoral spring. In: Biology of Chrysomelidae. Eds. Jolivet, P., Petitpierre, E., & Hsiao, T. H. Kluwar Academic, Dordrecht, Germany.

32. O'Brien, L. B., & Wilson, S. N. 1985. Planthopper systematics and external morphology. In: The Leafhoppers and Planthoppers, pp. 61–102. Eds. Nault, L. R., & Rodriguez, J. G. John Wiley & Sons, NY.

33. Hingston, Naturalist in the Guiana Forest.

34. Ibid.

35. Vickery, V. R., & Poinar, Jr., G. O. 1994. Crickets (Grylloptera: Grylloidea) in Dominican amber. Canadian Entomologist 126: 13–22.

36. Blatchley, W. S. 1920. Orthoptera of Northeastern America. Nature Publishing Co., Indianapolis.

37. De Vries, P. J., & Poinar, Jr., G. O. 1997. Ancient butterfly-ant symbiosis: Direct evidence from Dominican amber. Proc. Royal Soc. London B, 264: 1137–1140.

38. Smith, D. R., & Poinar, Jr., G. O. 1992. Sawflies (Hymenoptera: Argidae) from Dominican amber. Entomol. News 103: 117–124.

39. Santiago-Blay, J. A., Poinar, Jr., G. O., & Craig, P. R. 1996. Dominican and Mexican amber chrysomelids, with descriptions of two new species. In: Chrysomelidae Biology, Vol. 1, pp. 413–424. Eds. Jolivet, P.H.A., & Cox, M. L. SPB Academic Publishing, Amsterdam.

40. Poinar, Jr., G. O., Hess, R. T., & Tsitsipis, J. A. 1975. Ultrastructure of the bacterial symbiotes in the pharyngeal diverticulum of *Dacus oleae* (Gmelin) (Trypetidae: Diptera). Acta Zool. 56: 77–84.

41. Bright, D. E., & Poinar, Jr., G. O. 1994. Scolytidae and Platypodidae (Coleoptera) from Dominican Republic amber. Ann. Entomol. Soc. Amer. 87: 170–194.

42. Ahmadjian, V., & Paracer, S. 1986. Symbiosis. University Press of New England, Hanover, NH.

43. Beebe, W. 1925. Jungle days. G. P. Putman's Sons, New York.

44. Bright & Poinar, Scolytidae and Platypodidae.

45. Jaques, H. E. 1953. A long-lifed wood boring beetle. Proc. Iowa Acad. Sci. 25: 175.

46. Zeh, D. W., & Zeh, J. A. 1992. On the function of harlequin beetle-riding in the pseudoscorpion, *Cordylochernes scorpioides* (Pseudoscorpionida: Chernetidae). J. Arachnol. 20: 47–51.

47. Vaurie, P. 1957. *Atractocercus brasiliensis* in Cuba. Coleopterists Bull. 10: 86.

48. Simmonds, F. J. 1956. An investigation of the possibilities of biological control of *Melittomma insulare* Fairm. (Coleoptera, Lymexylonidae), a serious pest of coconut in the Seychelles. Bull. Entomol. Res. 47: 685–702.

49. Ahmadjian & Paracer, Symbiosis.

50. Dindal, D. L., ed. 1990. Soil Biology Guide. John Wiley & Sons, New York.

51. Poinar, Jr., G. O. 1983. The Natural History of Nematodes. Prentice Hall, Englewood Cliffs, NJ.

52. Holldobler, B., & Wilson, E. O. 1990. The Ants. Belknap Press, Cambridge, MA.

53. Krantz, G. W. 1978. A Manual of Acarology. (2d ed.). Oregon State University Book Store, Corvallis, OR.

54. Poinar, Natural History of Nematodes.

55. Nentwig, W., ed. 1987. Ecophysiology of Spiders. Springer-Verlag, Heidelberg.

56. Ibid.

57. Hingston, Naturalist in the Guiana Forest.

58. Cokendolpher, J. C., & Poinar, Jr., G. O. 1998. A new fossil harvestman from Dominican Republic amber (Opiliones: Samoidae: *Hummelinckiolus*). J. Arachnol. 26: 9-13.

59. Santiago-Blay, J., & Poinar, Jr., G. O. 1988. A fossil scorpion *Tityus geratus* new species (Scorpiones: Buthidae) from Dominican amber. Historical Biol. 1: 345–354.

60. Crawford, C. S. 1990. Scorpions, Solifugae and associated desert taxa. In: Soil Biology Guide, pp. 421–475. Ed. Dindal, D. L. John Wiley & Sons, New York.

61. Ibid.; Poinar, Jr., G. O., & Santiago-Blay, J. 1989. A fossil solpugid, *Happlodontus proterus*, new genus, new species (Arachnida: Solpugida) from Dominican amber. J. New York Entomol. Soc. 97: 125–132.

62. Schmallfuss, H. 1984. Two new species of the terrestrial isopod genus *Pseudarmadillo* from Dominican amber (Amber Collection Stuttgart: Crustacea, Isopoda, Pseudarmadillidae). Stuttgarter Beitr. Naturkunde (B), no. 102.

63. Bousfield, E. L., & Poinar, Jr., G. O. 1995. New terrestrial amphipod from Tertiary amber deposits of the Dominican Republic. J. Crustacean Biol. 15: 746–755.

64. Sturm, H., & Poinar, Jr., G. O. 1997. A new *Neomachilellus* species (Archaeognatha: Meinertellidae) from Miocene amber of the Dominican Republic and its phylogenetic relationships. Entomol. Gener. 22: 157–170.

65. Poinar, Jr.. G. O. 1988. *Zorotypus palaeus*, new species, a fossil Zoraptera (Insecta) in Dominican amber. J. New York Entomol. Soc. 96: 253–259.

66. Poinar, Jr., G. O., & Roth, B. 1991. Terrestrial snails (Gastropoda) in Dominican amber. Veliger 34: 253–258.

67. Santiago-Blay, J., & Poinar, Jr., G. O. 1992. Millipedes from Dominican amber, with the description of two new species (Diplopoda: Siphonophoridae) of *Siphonophora*. Ann. Entomol. Soc. Amer. 85: 363–369.

68. Mockford, E. L. 1986. A preliminary survey of Psocoptera from Tertiary amber of the Dominican Republic. Entomol. Soc. Amer. Conf. (Reno, Nevada, Dec. 7–11, 1986), p. 112.

69. Poinar, Jr., G. O., & Stange, L. A. 1996. A new antlion from Dominican amber (Neuroptera: Myrmeleontidae). Experientia 52: 383–386.

70. Poinar, Jr., G. O. 1996. Fossil velvet worms in Baltic and Dominican amber: Onychophoran evolution and biogeography. Science 273: 1370–1371.

71. Frank, J. H., & Lounibos, L. P., eds. 1983. Phytotelmata: Terrestrial Plants as Hosts for Aquatic Insect Communities. Plexus Publishing, Medford, NJ.

72. Benzing, D. H. 1990. Vascular Epiphytes. Cambridge University Press, New York.

73. Picado, C. 1913. Les bromeliacées epiphytes considerées comme milieu biologique. Bull. Scient. France, Belgium (7) 47: 216–360; Frank, J. H. 1983. Bromeliad phytotelmata and their biota, especially mosquitoes. In: Phytotelmata: Terrestrial Plants as Hosts for Aquatic Insect Communities, pp. 101–128. Eds. Frank, J. H., & Lounibos, L. P. Plexus Publishing, Medford, NJ.

74. Calvert, A. S., & Calvert, P. P. 1917. A Year of Costa Rican Natural History. Macmillan, New York.

75. Ibid.

76. Zavortink, T. J., & Poinar, Jr., G. O. 1999. A new species of *Anopheles (Nyssorhynchus)* from Dominican Amber (Diptera: Culicidae) (submitted).

77. Poinar, A Fossil Stalk-winged Damselfly.

78. Calvert & Calvert, A Year of Costa Rican Natural History.

79. Frank, Bromeliad phytotelmata.
80. Borror, D. J., Triplehorn, C. A., & Johnson, N. F. 1989. An Introduction to the Study of Insects. 6th ed. Saunders College Publishing, Philadelphia.
81. Stark, B. P., & Lentz, D. L. 1992. *Dominiperla antiqua*, the first stonefly from Dominican amber (Plecoptera: Perlidae). J. Kansas Entomol. Soc. 65: 93–96.
82. Wichard, W. 1989. Kocherfliegen des Dominikanischen Bernsteins. 7. Fossile Arten der Gattung *Cubanoptila* Sykora, 1973. Mitteiliungen Münch Entomol. Ges. 79: 91–100.
83. Frank, Bromeliad phytotelmata.
84. Borror et al., An Introduction.
85. Andersen, N. M., & Poinar, Jr., G. O. 1992. Phylogeny and classification of an extinct water strider genus (Hemiptera, Gerridae) from Dominican amber, with evidence of mate guarding in a fossil insect. Zeitsch. f. zoolog. Systematik und Evolutionsforsch. 30: 255–267.
86. Andersen, N. M., & Poinar, Jr., G. O. 1998. A marine water strider (Hemiptera: Veliidae) from Dominican amber. Entomol. Scand. 29: 1–9.
87. Schuh, R. T,. & Slater, J. A. 1995. True Bugs of the World (Hemiptera: Heteroptera). Cornell University Press, Ithaca, NY.
88. Borror et al., Introduction to the Study of Insects.
89. Holldobler & Wilson, The Ants.
90. Gotwald, Jr., W. H. 1995. Army Ants. Cornell University Press, Ithaca, NY.
91. Wilson, E. O. 1985. Ants of the Dominican amber (Hymenoptera: Formicidae). 2. The first fossil army ants. Psyche 92: 11–16.
92. Gotwald, Army Ants.
93. Ibid.
94. Baroni Urbani, C., & de Andrade, M. L. 1994. First description of fossil Dacetini ants with a critical analysis of the current classification of the tribe (Amber Collection Stuttgart: Hymenoptera, Formicidae. 6: Dacetini). Stuttgarter Beitr. Naturkunde (B), no. 198.
95. Caetano, F. H., & de Cruz-Landem, C. 1985. Presence of microorganisms in the alimentary canal of ants of the tribe Cephalotini (myricinae): Location and relationship with intestinal structures. Naturalia (Sao Paulo) 10: 37–47.
96. De Vries & Poinar, Ancient butterfly-ant symbiosis.
97. De Vries, P. 1991. Ecological and evolutionary patterns in riodinid butterflies. In: Ant-Plant Interactions, pp. 143-156. Ed. Huxley, C., & Cutler, D. F. Oxford University Press, New York.
98. Holldobler & Wilson, The Ants.
99. DeVries, Ecological and evolutionary patterns.
100. Nixon, G.E.J. 1951. The Association of Ants with Aphids and Coccids. Commonwealth Institute of Entomology, London.
101. Belt, T. 1874. The Naturalist in Nicaragua. John Murray, London.

102. Ibid; Kricher, J. C. 1989. A Neotropical Companion. Princeton University Press (2d ed., 1997).

103. Holldobler & Wilson, The Ants; Wheeler, W. M. 1907. The fungus-growing ants of North America. Bull. Amer. Mus. Nat. Hist. 23: 669–807.

104. Baroni Urbani, C. 1980. First description of fossil gardening ants (Amber Collection Stuttgart and Natural History Museum Basel: Hymenoptera: Formicidae 1: Attini). Stuttgarter Beitr. Naturkunde (B), no. 54.

105. Holldobler & Wilson, The Ants.

106. Creighton, W. S. 1950. The ants of North America. Bull. Mus. Comp. Zool., Cambridge, MA.

107. Ibid.

108. Baroni Urbani, C. 1980. The first fossil species of the Australian ant genus *Leptomyrmex* in amber from the Dominican Republic (Amber Collection Stuttgart: Hymenoptera, Formicidae. 3: Leptomyrmicini). Stuttgarter Beitr. Naturkunde (B), no. 62.

109. Donisthrope, H. St. J. K. 1927. The Guests of British Ants. George Routledge and Sons, London.

110. Nagel, P. 1997. New fossil paussids from Dominican amber with notes on the phylogenetic systematics of the paussine complex (Coleoptera: Carabidae). System. Entomol. 22: 345–362.

111. Poinar, Jr., G. O. 1991. *Praecoris dominicana* gen. n., sp. n. (Holoptilinae: Reduviidae: Hemiptera) from Dominican amber, with an interpretation of past behavior based on functional morphology. Entomol. Scand. 22: 193–199.

112. Ahmadjian & Paracer, Symbiosis.

113. Edwards, R., & Mill, A. E. 1986. Termites in Buildings: Their Biology and Control. Rentokil, East Grinstead, England.

114. Krishna, K., & Grimaldi, D. A. 1991. A new fossil species from Dominican amber of the living Australian termite genus *Mastotermes* (Isoptera: Mastotermitidae). Amer. Mus. Novitates 3021.

115. Cleveland, L. R. 1936. The wood-feeding roach *Cryptocercus*, its protozoa, and the symbiosis between protozoa and roach. Mem. Amer. Acad. Arts Sci. 17: 180–342.

116. Edwards & Mill, Termites in Buildings.

117. Hingston, Naturalist in the Guianan Forest.

118. Ibid.

119. Myers, J. G. 1932. Observations on the family Termitaphididae (Hemiptera-Heteroptera) with the description of a new species from Jamaica. Ann. Mag. Nat. Hist. 9: 366–373.

120. Wilson, E. O. 1971. The Insect Societies. Belknap Press, Cambridge, MA.

121. Michener, C. D., & Poinar, Jr. G. O. 1997. The known bee fauna of the Dominican amber. J. Kansas Entomol. Soc. 69: 353–361.

122. Ibid.

123. Poinar, Jr. G. O. 1999. Paleoeuglossa melissiflora gen. n., sp. n. (Euglossinae: Apidae), fossil orchid bees in Dominican amber. J. Kansas Entomol. Soc. 71: 29–34.

124. Dodson, C. H. 1975. Coevolution of orchids and bees. In: Coevolution of Animals and Plants, pp. 202–210. Ed. Gilbert, L. E., & Raven, P. H. University of Texas Press, Austin.

125. Michener, C. D., Mcginley, R. J., & Danforth, B. N. 1994. The bee genera of North and Central America (Hymenoptera: Apoidea). Smithsonian Institution Press, Washington, DC.

126. Michener & Poinar, Known bee fauna.

127. Roubik, D. W. 1989. Ecology and Natural History of Tropical Bees. Cambridge University Press, Cambridge, UK.

128. Ibid.

129. Poinar, Jr., G. O. 1992. Fossil evidence of resin utilization by insects. Biotropica 24: 466–468.

130. Usinger, R. L. 1958. Harzwanzen or "resin bugs" in Thailand. Pan-Pac. Entomol. 34: 52; Capriles, J. M., Santiago-Blay, J., & Poinar, Jr., G. O. 1993. *Apicrenus fossilis* gen. et sp. n. (Heteroptera: Reduviidae: Apiomerinae) from Dominican amber (Lower Oligocene-Upper Eocene). Entomol. Scand. 24: 139–142.

131. Poinar, Fossil evidence.

132. Michener & Poinar, Known bee fauna.

133. Wilson, Insect Societies; Gauld, I., & Bolton, B. Eds. 1988. The Hymenoptera. Oxford University Press, New York.

134. Prentice, M. A., & Poinar, Jr. G. O. 1993. Three species of *Trypoxylon* Latrielle from Dominican amber (Hymenoptera: Sphecidae). J. Kansas Entomol. Soc. 66: 280–291.

135. Coville, R. E. 1987. Spider hunting sphecid wasps. In: Ecophysiology of Spiders, pp. 319–327. Ed. Nentwig, W. Springer-Verlag, Heidelberg.

136. Ibid.

137. Wilson, Insect Societies.

138. Michener & Poinar, Known bee fauna.

139. Borror et al., Introduction to the Study of Insects.

140. La Salle, L., & Gauld, I. D. 1993. Hymenoptera and Biodiversity. CAB International, Wallingford, UK.

141. Ibid.

142. Gauld & Bolton, The Hymenoptera.

143. Goulet, H., & Huber, J. T., eds. 1993. Hymenoptera of the World: An Identification Guide to Families. Centre for Land and Biological Resources Research, Ottawa.

144. Francisco Alvarez, B. 1984. Studies on resistance of rice to *Sogatodes oryzicola* (Muir) and a parasitoid, *Haplogonatopus* sp., in Costa Rica. Master's thesis, Oregon State University, Corvallis.

145. Gauld & Bolton, The Hymenoptera.

146. Poinar, Jr., G. O., Hess, R., & Caltagirone, L. E. 1976. Virus-like particles in the calyx of *Phanerotoma flavitestacea* (Hymenoptera: Braconidae) and their transfer into host tissues. Acta Zool. 57: 161–165.

147. Krell, P. J. 1991. Polydnaviridae. In: Atlas of Invertebrate Viruses, pp. 321–338. Eds. Adams, J. R., & Bonami, J. R. CRC Press, Boca Raton, FL.

148. Zuparko, R. L., & Poinar, Jr., G. O. 1997. *Aivalykus dominicanus* (Hymenoptera: Braconidae), a new species from Dominican amber. Proc. Entomol. Soc. Wash. 99: 744–747.

149. Manley, D. G., & Poinar, Jr., G. O. 1999. A second species of fossil *Dasymutilla* (Hymenoptera: Mutillidae) from Dominican amber. Pan-Pac. Entomol. (in press).

150. Goulet & Huber, Hymenoptera of the World.

151. Gould & Bolton, the Hymenoptera.

152. Askew, R. R. 1971. Parasitic Insects. American Elsevier, NY.

153. Schlinger, E. I. 1987. The biology of Acroceridae (Diptera): True endoparasites of spiders. In: Ecophysiology of Spiders, pp. 319–327. Ed. Nentwig. W. Springer-Verlag, Heidelberg.

154. Clausen, C. P. 1962. Entomophagous Insects. Hafner, New York.

155. Perkins, R.C.L. 1905. Leaf-hoppers and their natural enemies (Pipunculidae). Hawaii Sugars Planters' Assoc. Expt. Sta. Bull. 1: 123–157.

156. Clausen, Entomophagous Insects.

157. Askew, Parasitic Insects; Clausen, Entomophagous Insects.

158. Askew, Parasitic Insects.

159. Ibid.; Clausen, Entomophagous Insects.

160. Roubik, Ecology and Natural History of Tropical Bees; Clausen, Entomophagous Insects.

161. Poinar, Naural History of Nematodes.

162. Poinar, Jr., G. O. 1968. *Hydromermis conopophaga* n. sp. parasitizing midges (Chironomidae) in California. Ann. Entomol. Soc. Amer. 61: 593–598.

163. Poinar, Jr., G. O., Acra, A., & Acra, F. 1994. Earliest fossil nematode (Mermithidae) in Cretaceous Lebanese amber. Fundam. Appl. Nematol. 17:475-477.

164. Poinar, Jr., G. O. 1984. First fossil record of parasitism by insect parasitic Tylenchida (Allantonematidae: Nematoda). J. Parasitol. 70: 306–308.

165. Poinar, Jr., G. O., Jaenike, J., & Dombeck, I. 1997. *Parasitylenchus nearcticus* sp. n. (Tylenchida: Allantonematidae) parasitizing *Drosophila* (Diptera: Drosophilidae) in North America. Fund. Appl. Nematol. 20: 187–190.

166. Poinar, Jr., G. O., & Thomas, G. M. 1984. Laboratory Guide to Insect Pathogens and Parasites. Plenum Press, New York.
167. Poinar, Jr., G. O., & Thomas, G. M. 1984. A fossil entomogenous fungus from Dominican amber. Experientia 40: 578–579; Poinar, Jr., G. O., & Thomas, G. M. 1982. An Entomophthorales from Dominican Amber. Mycologia 74: 332–334.
168. Poinar, Jr., G. O., & Hess, R. 1982. Ultrastructure of 40 million years old insect tissue. Science 215: 80–84.
169. Poinar, Jr., G. O., & Cannatella, D. C. 1987. An Upper Eocene frog from the Dominican Republic and its implications for Caribbean biogeography. Science 237: 1215–1216.
170. Scott, N. J. 1983. *Rhadinaea decorata* (Culebra). In: Costa Rican Natural History, p. 416. Ed. Jansen, D. H. University of Chicago Press, Chicago.
171. Kricher, A Neotropical Companion.
172. Frank, Bromeliad phytotelmata.
173. Rieppel, O. 1980. Green anole in Dominican amber. Nature 286: 486–487.
174. Bohme, W. 1984. Erstfund eines fossilen Kugelgingergeckos (Sauria: Gekkonidae: Sphaerodactylinae) aus Dominikanischem Bernstein (Oligozan von Hispaniola, Antillen). Salamandra 20: 212–220.
175. Smith, H.M. 1946. Handbook of Lizards. Comstock Publishing Company, Ithaca, NY.
176. Belt, Naturalist in Nicaragua.
177. Laybourne, R. C., Deedrick, D. W., and Hueber, F. M. 1994. Feather in amber is earliest New World fossil of Picidae. Wilson Bull. 106: 18–25.
178. Poinar, Jr., G. O. 1988. Hair in Dominican amber: Evidence for Tertiary land mammals in the Antilles. Experientia 44: 88–89.
179. Bolay, The Dominican Rebublic.
180. McClung, R. M. 1981. Vanishing Wildlife of Latin America. William Morrow, New York.
181. Reddish, P. 1996. Treasure islands. BBC Wildlife Magazine 14: 18–26.
182. MacPhee, R.D.E., & Grimaldi, D. A. 1996. Mammal bones in Dominican amber. Nature 380: 489–490.
183. Poinar, Jr., G. O. 1995. Fleas (Insecta: Siphonaptera) in Dominican amber. Med. Sci. Res. 23: 789.
184. Lane, R. S., & Poinar, Jr., G. O. 1986. First fossil tick (Acari: Ixodidae) in New World amber. Int. J. Acarology 12: 75–78.
185. Poinar, Jr., G. O. 1995. First fossil soft ticks, *Ornithodoros antiquus* n. sp. (Acari: Argasidae) in Dominican amber, with evidence of their mammalian host. Experientia 51: 384–387.
186. Poinar, Jr., G. O., & Columbus, J. T. 1992. Adhesive grass spikelet with mammalian hair in Dominican amber: First fossil evidence of epizoochory. Experientia 48: 906–908.

187. Forsyth, A., & Miyata, K. 1984. Tropical Nature. Charles Scribner's Sons, New York.

188. Borror et al., Introduction to the Study of Insects.

Chapter 3. Reconstruction of the Amber Forest

1. Holdridge, L. R., et al. 1971. Forest Environments in Tropical Life Zones: A Pilot Study. Pergamon Press, Oxford.

2. Simpson, G. G. 1956. Zoogeography of West Indian land mammals. Amer. Mus. Novitates 1759.

3. Domning, D.P., Emry, R.J., Portell, R.W., Donovan, S.K., & Schindler, K.S. 1997. Oldest West Indian land mammal; Rhinocertotoid ungulate from the Eocene of Jamaica. J. Vert. Paleontol. 17: 638–541.

4. Trewartha, G.T. 1968. An Introduction to Climate. McGraw-Hill, New York.

5. Macdonald, D. ed. 1984. The Encyclopedia of Mammals. Facts-on-File Publications, New York.

6. MacPhee, R.D.E., & Iturralde-Vincent, M. A. 1994. First Tertiary land mammal from Greater Antilles: An early Miocene sloth (Xenarthra, Megalonychidae) from Cuba. Amer. Mus. Novitates 3094.

7. Morgan, G. S., & Woods, C. A. 1986. Extinction and the zoogeography of West Indian land mammals. Biol. J. Linnean Soc. 28: 167–203.

8. Poinar, Hair in Dominican amber.

9. Miller, Jr., G. S. 1929. Mammals eaten by Indians, owls, and Spaniards in the coast region of the Dominican Republic. Smithsonian Misc. Coll. 82: 1–18.

10. Woods, C. A. 1989. The biogeography of West Indian rodents. In: Biogeography of the West Indies: Past, Present and Future, pp. 741–798. Ed. Woods, C. A. Sandhill Crane Press, Gainesville, FL.

11. Williams, E. E,. & Koopman, K. F. 1952. West Indian fossil monkeys. Amer. Mus. Novitates 1546.

12. MacPhee, R.D.E., & Woods, C. A. 1982. A new fossil Cebine from Hispaniola. Amer. J. Phys. Anthropol. 58: 419–436.

13. Bolay, The Dominican Republic.

14. Baker, R. J., & Genoways, H. H. 1978. Zoogeography of Antillean bats. In: Zoogeography in the Caribbean. Ed. Gill, F. B. Academy of Natural Sciences of Philadelphia, Special Publication 13: 53–97.

15. Domning et al., Oldest West Indian land mammal.

16. Romer, A. S. 1966. Vertebrate Paleontology. University of Chicago Press, Chicago.

17. Poinar & Columbus, Adhesive grass spikelet.

18. Morgan & Woods, Extinction and zoogeography.

19. MacPhee, R.D.E., & Wyss, A. R. 1990. Oligo-Miocene vertebrates from Puerto Rico, with a catalog of localities. Amer. Mus. Novitates 2965.
20. Schwartz, A. 1978. Some aspects of the herpetogeography of the West Indies. In: Zoogeography in the Caribbean. Ed. Gill, F. B. Academy of Natural Sciences of Philadelphia, Special Publication 13: 31–51.
21. Ibid.; Reddish, Treasure islands.
22. Schwartz, Some aspects.
23. Bond, J. 1985. Birds of the West Indies. Houghton Mifflin, Boston.
24. Olsen, S. L. 1978. A paleontological perspective of West Indian birds and mammals. In: Zoogeography in the Caribean. Ed. Gill, F. B. Academy of Natural Sciences of Philadelphia, Special Publication 13: 99–117.
25. Ibid.
26. Bond, Birds of the West Indies.
27. Cockerell, T.D.A. 1924. A fossil cichlid from the Republic of Haiti. Proc. U.S. Nat. Mus. 63: 1–2.
28. Bolay, The Dominican Republic.
29. Varona, L. S. 1964. Un dugongido del Mioceno de Cuba (Mammalia: Sirenia). Memoirs, Sociedad de Ciencias naturales La Salle, Caracas, Venezuela 32.
30. Iturralde-Vincent & MacPhee, Age and paleogeographic origin of Dominican Amber. Science 273: 1850–1852.
31. Schlee, Das Bernstein-Kabinett.
32. Boucot, Evolutionary Paleobiology.
33. Holdridge et al., Forest Environments.
34. Croat, T. B. 1978. Flora of Barro Colorado Island. Stanford University Press, Stanford, CA.
35. Bolay, The Dominican Republic.
36. Croat, Flora of Barro Colorado.
37. Castner, J. L. 1990. Rainforests: A Guide to Research and Tourist Facilities at Selected Tropical Forest Sites in Central and South America. Feline Press, Gainesville, FL.
38. Doyen, J. T., & Poinar, Jr., G. O. 1994. Tenebrionidae from Dominican amber (Coleoptera). Entomol. Scand. 25: 27–51.
39. Creighton, The Ants of North America.
40. Hartshorn, G. S. 1983. Plants. Introduction. In: Costa Rican Natural History. Ed. Janzen, D. H. University of Chicago Press, Chicago.
41. Myers, N. 1979. The Sinking Ark. Pergamon Press, Oxford.
42. Jacobs, Tropical Rain Forest; Bolay, the Dominican Republic.
43. Michener & Poinar, Known bee fauna; Poinar, *Paleoeuglossa melissiflora.*
44. Purchas, S. 1657. A theatre of Politicall flying-insects. London. (Cited by C. D. Michener in Melissa, no. 8, August 1994, p. 9.)
45. Roubik, Ecology and Natural History of Tropical Bees.

46. Stanley, S. M. 1984. Marine mass extinctions: A dominant role for temperature. In: Extinctions, 69-117. Ed. Nitecki, M. H. University of Chicago Press, Chicago. Stanley, S. M. 1987. Extinction. Scientific American Library, New York. Allmon, W. D., Emslie, S. D., Jones, D. S., & Morgan, G. S. 1996. Late Neogene oceanographic change along Florida's west coast: Evidence and mechanisms. J. Geology 104: 143–162. Allmon, W. D., Rosenberg, G., Portell, R. W., & Schindler, K. S. 1993. Diversity of Atlantic coastal Plain Molluscs since the Pliocene. Science 260: 1626–1629.

47. Stanley, Marine mass extinctions; Allman et al., Late Neogene change.

48. Avise, J. C. 1994. Molecular markers, natural history and evolution. Chapman & Hall, New York.

49. Stanley, Extinction.

50. Ibid.

51. Michener, C. D. 1979. Biogeography of the bees. Ann. Missouri Bot. Garden 66: 277–347.

52. Trewartha, Introduction to Climate.

53. Glick, P. A. 1939. The distribution of insects, spiders and mites in the air. USDA Technical Bull. 673: 1–151.

54. Poinar, Life in Amber.

55. Watson, J.A.L., & Gay, F. J. 1991. Isoptera. In: The Insects of Australia, pp. 330–347. Ed. Naumann, I. D. Cornell University Press, Ithaca, NY.

Chapter 4. Amber Today

1. Poinar, Jr., G. O. 1996. Older and wiser. Lapidary Journal 49: 52–56.

2. Poinar, Jr., G. O. 1982. Amber; true or false? Gems and Minerals 534: 80–84.

3. Higuchi, R., Bowman, B., Freeberger, M. Ryder, O. A., & Wilson, A. C. 1984. DNA sequence from the quagga, an extinct member of the horse family. Nature 312: 282–284.

4. Poinar & Hess (1982), Ultrastructure of 40-million years old insect.

5. Cano, R. J., Poinar, H., & Poinar, Jr., G. O. 1992. Isolation and partial characterization of DNA from the bee *Proplebeia dominicana* (Apidae: Hymenoptera) in 25–40 million-year-old amber. Med. Sci. Res. 20: 249–251; Cano, R. J., Poinar, H. N., Roubik, D., & Poinar, Jr., G. O. 1992. Enzymatic amplification and nucleotide sequencing of portions of the 18s rRNA gene of the bee *Proplebeia dominicana* (Apidae: Hymenoptera) isolated from 25–40 million-year-old Dominican amber. Med. Sci. Res. 20: 619–622.

6. DeSalle, R., Gatesy, J., Wheeler, W., & Grimaldi, D. 1992. DNA sequences from a fossil termite in Oligo-Miocene amber and phylogenetic implications. Science 257: 1880–1882.

7. Poinar, H. N., Cano, R. J., & Poinar, Jr., G. O. 1994. DNA from an extinct plant. Nature 363: 677.

8. Cano, R. J., Poinar, H. N., Pieniazek, J. J., Acra, A., & Poinar, Jr., G. O. 1993. Amplification and sequencing of DNA from a 120–135 million year old weevil. Nature 363: 536–538.

9. Poinar, H. N., Melzer, R. R., & Poinar, Jr., G. O. 1996. Ultrastructure of 30–40 million year old leaflets from Dominican amber (*Hymenaea protera*, Fabaceae: Angiospermae). Experientia 52: 387–390.

10. Wang, X. S., Poinar, H. N., Poinar, Jr., G. O., & Bada, J. L. 1995. Amino acids in the amber matrix and in entombed insects. In: Amber, Resinite and Fossil Resins, pp. 255–262. Eds. Andersen, K. B., & Crelling, J. C. American Chemical Society, Washington, DC; Poinar, H. N., Hoss, M., Bada, J. L., & Paabo, S. 1996. Amino Acid Racemization and the preservation of Ancient DNA. Science 272: 864-866.

Index

(entries in italics indicate figures)

Tingidae, *54*, 59, *59*
Tipulidae, *168*
tissue: extraction of, 195; preservation of, 193; protection of, 196
todies, 176
tortricids (Tortricidae), 55
traces, of mammals, 160
traps, ant or pit, 85, 129
treehoppers, 50, *51*, *54*, 103
trees, in amber forest, 13–24, *181*; algarrobo, xiii; emergent, 11; acacia, *22*; kauri, 192
Trichilia (*see* souca), 19, 21, 182; flowers, 21
trichogrammatid (Trichogrammatidae), 134
trichomes (*see* plant hairs), 29
Trichopseniidae, *xiv*
Tridactylidae, *96*
Trigonidiidae (bush cricket), *xiv*, 55, *55*, *162*
triungulin, 141, *149*
trogan, 176
tropical forest, structure of, 11; characterization of, 11–12; categories of, 179; rainfall of, 169, 179
Trypoxylon (wasp), 126, *126*
Trypoxylon eucharis (wasp), *126*
tunnels, of beetles, 61, *62*
turtles, 175
twisted-wing parasites, 145, *145*, *146*, 147; male attracted to odor of female, 145
Typhlophidae, *158*

uloborid (Uloboridae), *75*
ultrastructure, of amber fossils, 194; of cells in amber, 196
understory, 11, 182; plants of, 23
ungulates, in Antilles, 170, 174

Varicella (snail), *82*
vegetation, strata of, 11; semideciduous, 179
Velidae, *95*
velvet ants, 139, *139*
velvet mite, *xiv*
velvet worms, 86, *86*, 87, 91, *141*; oral papillae, 87; slime secretion, 87; attacking prey, *159*

venom: of velvet ants, 139; of wasps, 127
vertebrates: direct evidence of, 151–161; indirect evidence of, 161
Vespidae, 125, 127, *127*
vicariance, 170, 187
vines, 24, *181*, 182; evidence of, *24*; herbaceous, 11
Virola (*see* sebo), 18; flower, *19*
viruses: as means of defense, 138; transmission of, 167

walking sticks, 59; egg of, *xiv*
Walterianella (beetle), *43*, *45*
wasps, 58; as parasitoids, 133; as prey, 148; attacking caterpillar, *101*, 102; braconid, 137, 138, *138*, *139*; chalcid, *101*, *134*; cuckoo, 135, *136*; defenses of cuckoo, 135; ensign, 137; fig, 30–34, *30*, *33*, *39*; ichneumonid, 137, 138; paper, 125, 127, *127*, *128*; pemphredines, 125; protecting nests, 99; pupa of, 99; reaction to army ants, 99; roach, 125; sac, 135, *136*; social, 99, 125; sphecids, 125; spider, 126, *126*, 139, *140*
water striders, 94, *94*, 95, *95*; broad-shouldered, 95, *95*
wax production, 48; from planthopper tail, 50, *50*
web spinner, *xiv*
webs, spider, 60, *73*
weevils, 44, *44*, 60; DNA from, 194; polydrusine, *44*, *45*; zygopine, *45*, 46, *46*
whipscorpions, tail-less, 78, *78*
winds, prevailing in Hispaniola, 178
windscorpion, 78, *78*
wood lice, *77*, 79, *79*
woodpecker, 158, *158*, 160
worker, replete ant as evidence of dry periods, 108, *109*, 179
worms (*see also* nematodes, hairworms), white, 92

Xestocephalus (leafhopper), 50, *54*
Xylaria (fungus), 40, 41, *37*, *42*
Xylariaceae, *37*, *42*

zebra, 194
zorapteran (Zoraptera), 81, *81*, 82, *141*
Zorotypus palaeus, 81, *81*